Conference Proceedings of the Society for Experimental Mechanics Series

Series Editor

Kristin B. Zimmerman, Ph.D.
Society for Experimental Mechanics, Inc.,
Bethel, CT, USA

More information about this series at http://www.springer.com/series/8922

Bonnie Antoun
Editor

Challenges in Mechanics of Time Dependent Materials, Volume 2

Proceedings of the 2015 Annual Conference on Experimental and Applied Mechanics

Springer

Editor
Bonnie Antoun
Sandia National Laboratories
Livermore, CA, USA

ISSN 2191-5644 ISSN 2191-5652 (electronic)
Conference Proceedings of the Society for Experimental Mechanics Series
ISBN 978-3-319-22442-8 ISBN 978-3-319-22443-5 (eBook)
DOI 10.1007/978-3-319-22443-5

Library of Congress Control Number: 2015952466

Springer Cham Heidelberg New York Dordrecht London

Printed on acid-free paper

Springer International Publishing AG Switzerland is part of Springer Science+Business Media (www.springer.com)

Preface

Challenges in Mechanics of Time-Dependent Materials represents one of nine volumes of technical papers presented at the SEM 2015 SEM Annual Conference and Exposition on Experimental and Applied Mechanics organized by the Society for Experimental Mechanics and held in Costa Mesa, CA, June 8–11, 2015. The complete Proceedings also includes volumes on: Dynamic Behavior of Materials; Advancement of Optical Methods in Experimental Mechanics; Experimental and Applied Mechanics; MEMS and Nanotechnology; Mechanics of Biological Systems and Materials; Mechanics of Composite and Multifunctional Materials; Fracture, Fatigue, Failure and Damage Evolution; and Residual Stress, Thermomechanics and Infrared Imaging, Hybrid Techniques and Inverse Problems.

Each collection presents early findings from experimental and computational investigations on an important area within Experimental Mechanics, the Mechanics of Time-Dependent Materials being one of these areas.

This track was organized to address constitutive, time (or rate)-dependent constitutive, and fracture/failure behavior of a broad range of materials systems, including prominent research in both experimental and applied mechanics. Papers concentrating on both modeling and experimental aspects of Time-Dependent Materials are included.

The track organizers thank the presenters, authors, and session chairs for their participation and contribution to this track. The support and assistance from the SEM staff is also greatly appreciated.

Livermore, CA, USA

Bonnie Antoun

Contents

Chapter 1
Thermal Degradation of Extension Springs

Brian T. Werner, Bonnie R. Antoun, and George B. Sartor

Abstract Extension springs are used to apply a constant force at a set displacement in a wide variety of components. When subjected to an abnormal thermal event, such as in a fire, the load carrying capacity of these springs can degrade. In this study, relaxation tests were conducted on extension springs where the heating rate and dwell temperature were varied to investigate the reduction in force provided by the springs. Two commonly used spring material types were tested, 304 stainless steel and Elgiloy, a cobalt-chrome-nickel alloy. Challenges associated with obtaining accurate spring response to an abnormal thermal event are discussed. The resulting data can be used to help develop and test models for thermally activated creep in springs and to provide designers with recommendations to help ensure the reliability of the springs for the duration of the thermal event.

Keywords Time-dependent response • Thermomechanical behavior • Creep • Relaxation

1.1 Introduction

Extension springs are widely used in mechanical components and assemblies to provide a constant force at a set displacement. Such components may be subject to abnormal thermal transients well beyond normal operating conditions, such as a fire. Current design methodologies are not equipped to account for such extreme cases. This produces significant uncertainty in the performance of extension springs when they are subjected to high temperatures. Multiple mechanisms can contribute to the relaxation of extension springs within a component: thermal expansion of supports and the spring itself, reduction in the elastic or shear modulus, microstructural changes, and plastic creep. Assessing the effect spring geometry and material choice can have on each of these factors could either extend the time to failure of the spring or prevent failure altogether.

Dykhuizen and Robinho performed an extensive experimental study on the performance of an Elgiloy extension spring to a variety of heating profiles in order to develop a creep model that can predict relaxation behavior [1]. While the work is detailed and the model's performance is excellent, it is unclear whether this model is purely empirical where a similarly detailed study would have to be performed on each spring material and geometry to produce an accurate relaxation model. This study hopes to replicate the ability to collect excellent data which can be applied in a finite element model to allow designers the flexibility to change spring geometry while maintaining high temperature performance without the need of another experimental study.

In this study, two different springs, of material and geometry, were tested under relaxation. One spring was made of 304 stainless steel while the other was made of a cobalt-nickel-chrome alloy whose trade name is Elgiloy. A clamshell furnace was used to produce a controlled temperature ramp rate to a desired temperature which was then held for a predetermined amount of time. The load on the spring was monitored during the ramp, dwell, and cool down.

1.2 Experimental Procedures

Two springs were investigated in this study, one made of Elgiloy and one made of 304 stainless steel (Fig. 1.1). While they do not have the same geometry or carry the same service load, a comparison between their responses can be made on a purely qualitative basis. Each spring was designed to be used at a specified displacement to provide a constant force. It can be

B.T. Werner (✉) • B.R. Antoun • G.B. Sartor
Sandia National Laboratories, Livermore, CA 94550, USA
e-mail: btwerne@sandia.gov; brantou@sandia.gov

B. Antoun (ed.), *Challenges in Mechanics of Time Dependent Materials, Volume 2*,
Conference Proceedings of the Society for Experimental Mechanics Series, DOI 10.1007/978-3-319-22443-5_1

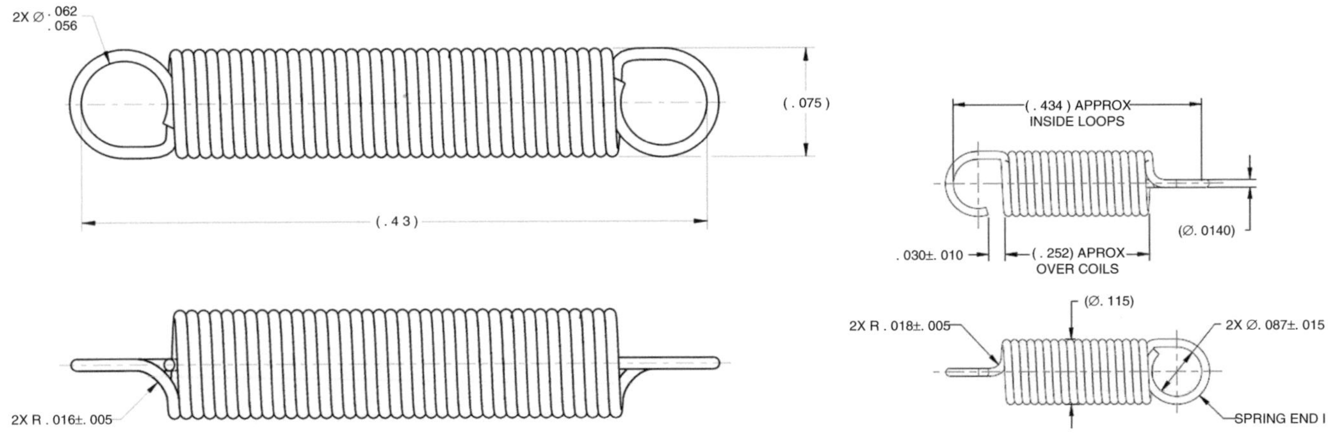

Fig. 1.1 Spring dimensions for (**a**) Elgiloy and (**b**) 304 stainless steel

difficult to determine where its relaxed state is so the springs were extended until the specified force was reached (0.294 lbs for the Elgiloy spring and 0.85 lbs for the 304 stainless steel spring).

A simple spring model was used to separate each of the different displacement modes during the test in order to try to isolate the relaxation that is due to microstructural changes and thermally activated creep in the spring.

$$F = k(T)\left[\delta_i - \delta_f(T) - \delta_s(T) - \delta_c\right] \tag{1.1}$$

F = *Force on spring*
k(*T*) = *Spring constant*
δi = *Initial spring displacement*
δf(*T*) = *Thermal expansion of fixture*
δa(*T*) = *Thermal expansion of spring*
δc = *Spring extension due to creep*

Before the relaxation due to creep can be quantified, the other sources of relaxation must be investigated. The reduction in spring constant and thermal expansion of the spring due to temperature is inherent due to the spring's geometry and material but through effective experimental design, the impact of thermal expansion in the fixturing can be limited.

The relaxation tests were performed on an MTS Bionix system. Due to the small geometry and loads of the springs being investigated, a 1.1 lb Interface load cell was used to track the force on the spring. In previous studies the springs were heated with the use of an induction heater and susceptor. In order to provide more control over the heating and a simple procedure to set up a spring, a clamshell furnace with a PID controller was chosen for this study. Passing into the furnace are two 5/16″ quartz rods with nichrome wire hooks that are used to attach to the spring ends. The springs are too small to weld thermocouples to them so two type K thermocouple probes are also tied in place just above the spring to record the air temperature and provide feedback to the PID controller. This may provide some error between the air temperature and the spring temperature but since the wire that the springs are made of is thin this error is most likely small. Images of the furnace, spring mounting, and complete setup, including load cell, are shown in Fig. 1.2.

One concern in using a clamshell furnace is temperature variation along its axial length. This was tested by keeping the PID controller's thermocouple mounted at the center of the furnace so the temperature was kept constant and the temperature of a second thermocouple was recorded as it was plunged from outside of the furnace towards the center. Figure 1.3 shows the temperature of this second thermocouple as it was moved from the center of the furnace outwards. The temperature is fairly constant until the thermocouple is 0.75″ from the center. This suggests that there is a 1.5″ region within the furnace of constant temperature. As long as the spring is within that 1.5″ region, it will be of constant temperature.

Another point of concern is the impact that the expansion of the fixture has on the results. When the fixture expands, the load in the spring drops since the distance between its supports closes. This study is not interested in relaxation due to changes in support so it is important to minimize the thermal expansion of the fixture. Quartz rods are used as pull rods in the furnace. Quartz has a very low coefficient of thermal expansion so this is the best choice. Most of the expansion seen in

Fig. 1.2 Fixturing and clamshell furnace

Fig. 1.3 Temperature variation within clamshell furnace

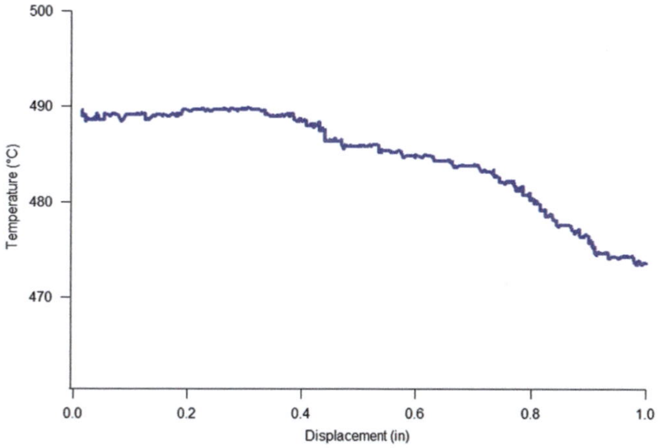

the fixture arises from the use of nichrome wire as a hanger material. In Fig. 1.4 the two hangers were passed through one another and a creep test was run at 1 lb. The temperature was ramped at 25 °C/min to 600 °C, held for 10 min, cooled, then reheated at 25 °C/min to 700 °C and held for 10 min while the displacement was measured using a laser extensometer.

There is a significant difference between the first loading and the second loading. This is most likely due to relaxation of the wire at high temperature as it conforms to its most taut position. Subsequent thermal cycles do not include this displacement. What is of more concern is that when the hanger wire reaches the dwell temperature, it continues to creep. Since the study hopes to isolate the creep present within the spring, any creep in the fixturing will cloud the issue. It is suggested that materials other than nichrome be investigated. One such material is invar that has a very low coefficient of thermal expansion up to around 400 °C. Whether it remains low up to 700 °C and does not creep is another question that needs to be answered. The best solution however would be to machine a hook into the end of the quartz rod that the spring can sit on. This is currently being attempted.

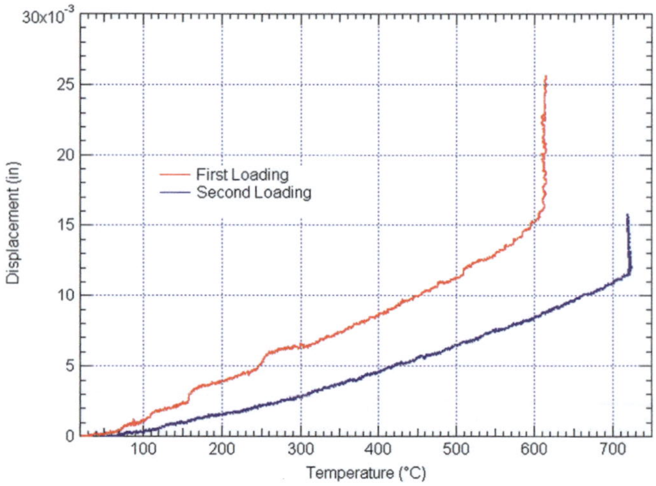

Fig. 1.4 Thermal expansion and creep of fixturing

1.3 Results

In this study the effect of extreme temperature on the load carrying performance of two different extension springs was investigated. In order to better understand the spring's response to high temperature, each of the parameters in (1.1) needed to be quantified. At this point the spring is being treated like a structure. Converting the measured force into a resultant displacement requires an accurate spring constant. This spring constant may change as the temperature rises due to microstructural effects. The spring was heated in 50 °C increments between 200 and 700 °C and at each temperature it was cycled 0.060 in. three times to determine the spring constant (Fig. 1.5). As the temperature increases, the spring constant decreases. A linear fit was made to the data to determine the spring constant. It can even be seen that the bands of data begin to spread due to creep at higher temperatures. This does not affect the spring constant calculations as the data bands include an equal number of points collected in both loading and unloading.

The resulting spring constants are shown in Fig. 1.6. There is a linear decrease in stiffness until 500 °C for the Elgiloy spring and then it drops off significantly. This trend matches that reported by Dykhuizen and Robinho [1] for an Elgiloy spring.

For the spring relaxation tests, a typical temperature and load profile is shown in Fig. 1.7 for the Elgiloy spring. The temperature ramp is at 5 °C/min, the dwell temperature is 600 °C, and the dwell time is 30 min. The initial load drop is dominated by thermal expansion of the fixture and the spring coil itself. As the temperature gets higher, thermal creep begins to set in and that accounts for the steep drop in load. Once the furnace is turned off, the load begins to rebound as the spring and fixture cool, applying more displacement on the spring itself.

Once the spring is forced to endure such an extreme temperature cycle there is a question of the degree of damage to its performance. Figure 1.8 shows that the spring constant does not change from before and after the temperature cycle. So while there is significant plastic creep in the spring, shown by the drop in load at the same displacement, its stiffness remains unaffected.

The thermal expansion from the fixturing was covered in the previous section so now it is necessary to correct for the expansion of the spring coil itself. Since both springs are a closed coil, as the temperature rises the coil's expansion can produce a significant relaxation in the load. Since the goal of the study is to isolate the relaxation due to creep, it is important to correct for this expansion as well. After the data has been corrected for the fixture expansion, the load relaxation is plotted against the temperature during the test (Fig. 1.9). The contribution from spring creep is assumed to be very small at temperature less than 200 °C, so the portion of the curve from room temperature to 200 °C is taken to represent the degree of relaxation in the spring due to a change in temperature. This portion of the curve is linear and it is very close to the slope seen in the cool down region. For this reason, the initial relaxation is extrapolated to cover the full temperature range of the test. This may provide some error as the spring material may not have a consistent coefficient of thermal expansion over the full temperature range and deformations from the thermal creep may localize and provide non-axial expansions or contractions of the spring coil itself. However without further study this assumption is a best case scenario at this stage. Iterative modeling with finite element may improve upon this assumption.

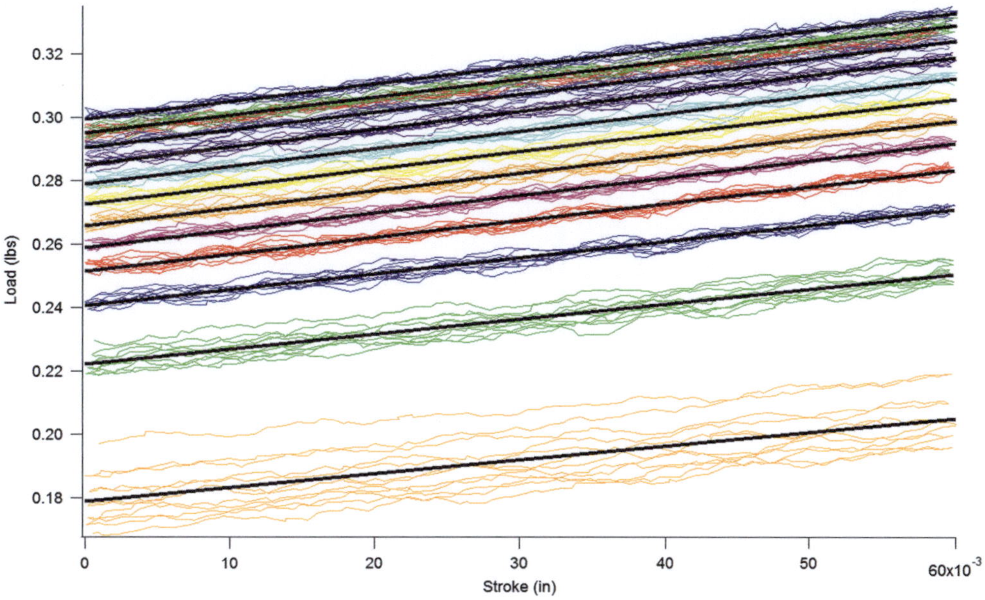

Fig. 1.5 Load–displacement curves for Elgiloy spring at different temperatures

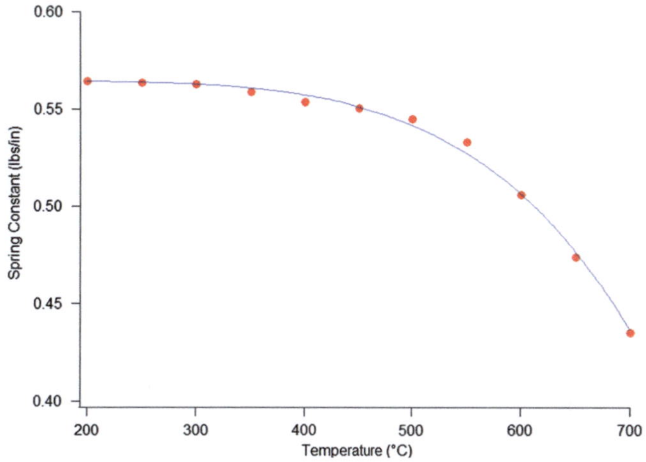

Fig. 1.6 Spring constant for Elgiloy spring at various temperatures

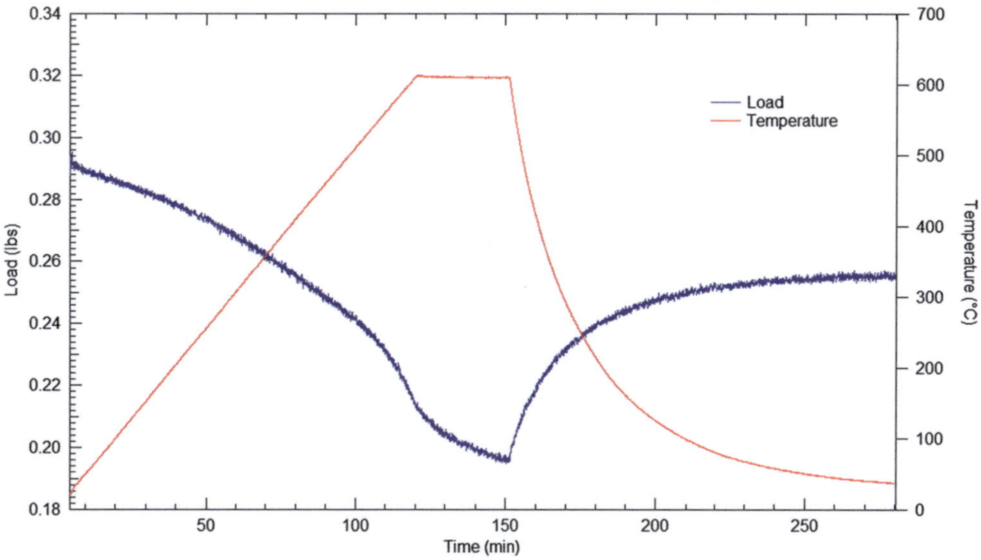

Fig. 1.7 Raw data for spring relaxation testing

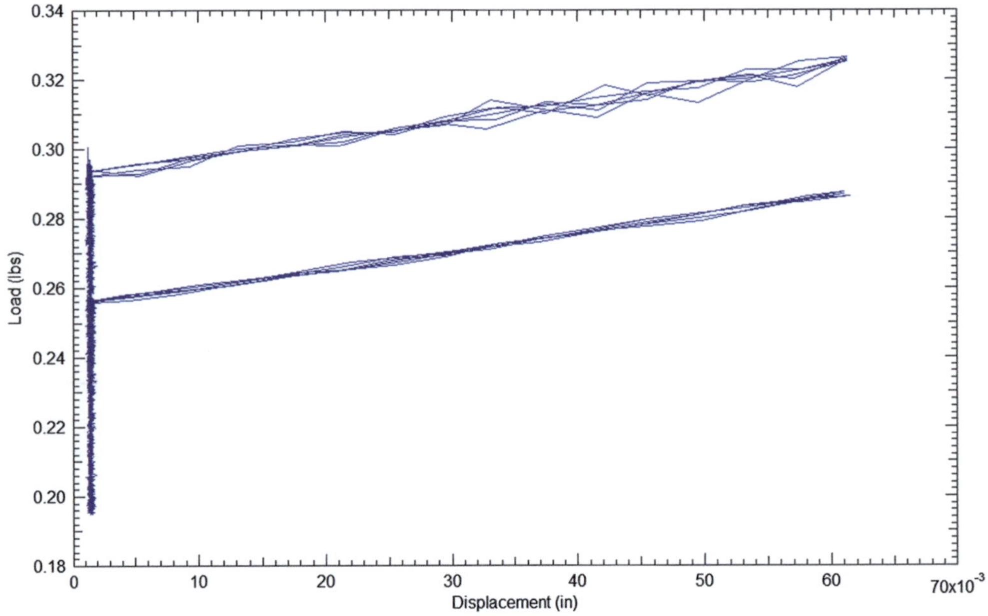

Fig. 1.8 Load–displacement curve for Elgiloy spring pre- and post-test

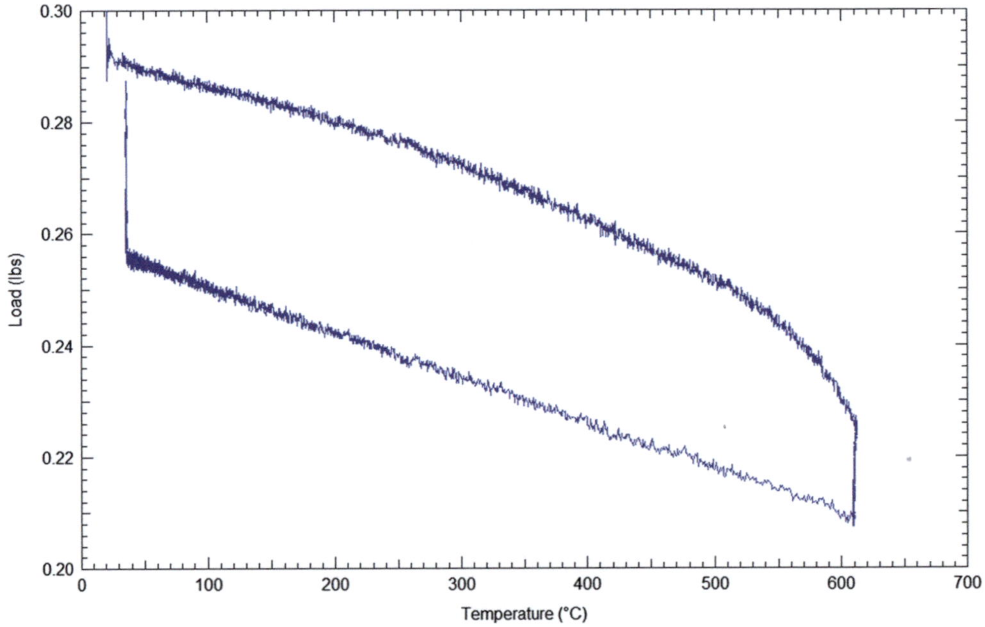

Fig. 1.9 Load-temperature curve for Elgiloy spring relaxation test

By correcting for the thermal expansion in the fixture and the spring, the effect of thermal creep begins to emerge. In Fig. 1.10 the effect of each correction on an Elgiloy spring is shown. The fixture itself provides a small amount of error as seen with the small difference between the red and blue curves. The larger source of load relaxation is due to the expansion of the spring itself. Within the component, this relaxation is present and cannot be ignored, but it is mostly recoverable when the spring cools so it does not represent plastic damage to the spring. Once it is removed the S-curve from thermal creep emerges where it does not begin affecting spring performance until well into the test. A small rebound in load is still present once the cooling begins. This is most likely due to either error in the fixture and spring expansion assumptions, such as nonlinearity with temperature, or as the result of plastic damage to the spring itself during the temperature cycle. The resulting thermal creep data is now isolated so a creep model can be fit.

The Elgiloy springs showed very little damage from thermal creep even up to 500 °C. For this reason the springs were test at 500, 600, and 700 °C (Fig. 1.11). The 304 stainless steel springs showed thermal creep damage at much lower tem-

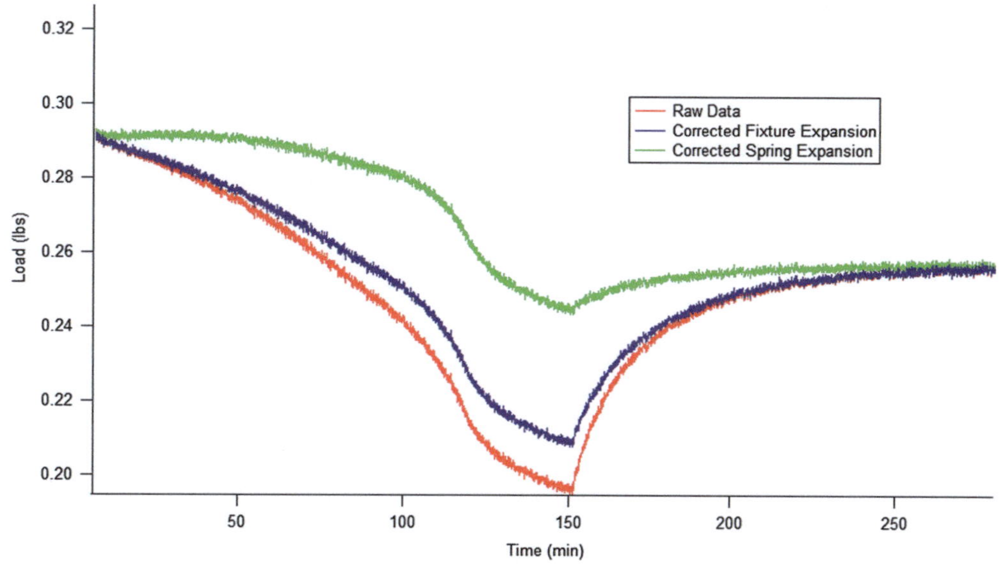

Fig. 1.10 Correction to remove fixture and spring thermal expansion from raw data

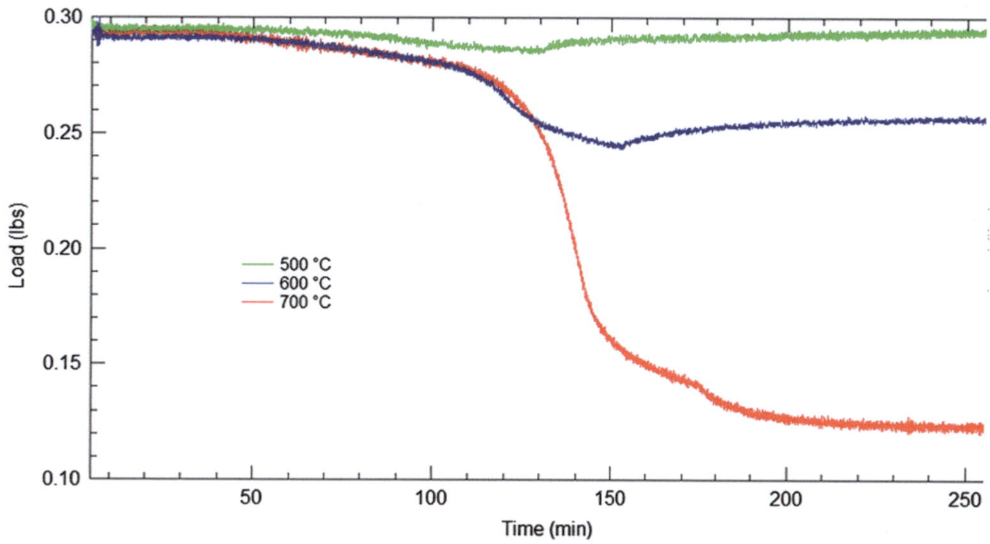

Fig. 1.11 Elgiloy spring relaxation at 500, 600, and 700 °C

peratures so they were only tested at 400, 500, 600 °C (Fig. 1.12). At 600 °C the steel spring relaxed to around 10 % of its initial load value so it did not seem necessary to test it at 700 °C. As expected all of the load relaxation curves appear to track closely to one another until a higher temperature is reached. Once the thermal creep begins to dominate the relaxation mode, the load drops off steeply until the dwell is reached at which point it is slowed to an almost linear slope. This information is quite useful when judging the danger of a fast acting fire. The temperature dwell point seems to have a much larger impact on the load carrying capacity of the spring than the duration of the elevated temperature.

The Elgiloy springs appear to perform much better than the 304 stainless steel springs. Not only does the onset of thermal creep occur later in the test, which corresponds to a higher temperature, its performance at higher temperatures is much better. Even at 700 °C, the Elgiloy spring still maintains over one third of its initial load while at 600 °C the 304 stainless steel spring can only hold about 10 % of its initial load.

Comparing the two springs at 600 °C may not be entirely valid due to their different geometries and initial loads and displacements, but, after normalizing for initial load, the results are still quite striking (Fig. 1.13). In the Elgiloy spring, thermal creep doesn't appear until the temperature has reached around 400 °C while it shows up at around half that temperature in the 304 stainless steel springs. This lower onset temperature contributes to the steep decline in performance for the stainless steel spring.

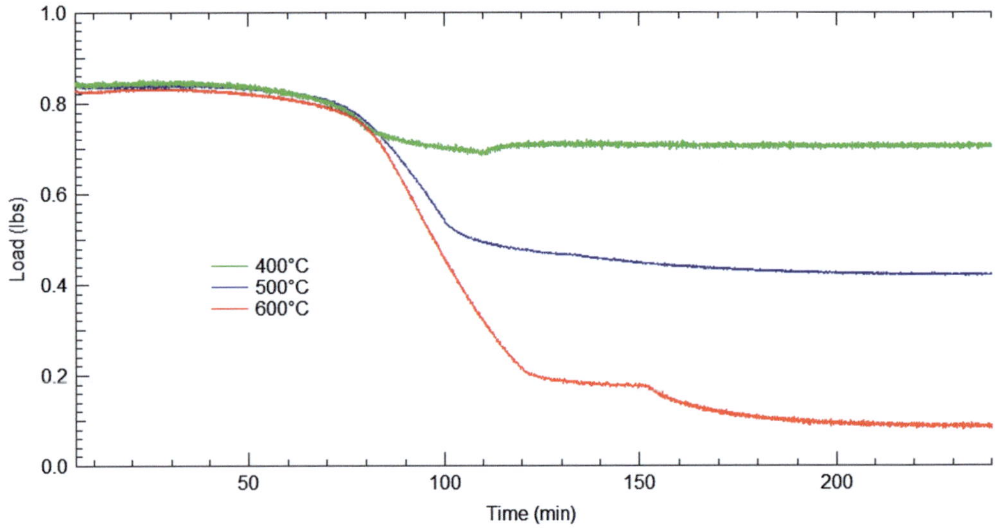

Fig. 1.12 304 stainless steel spring relaxation at 400, 500, and 600 °C

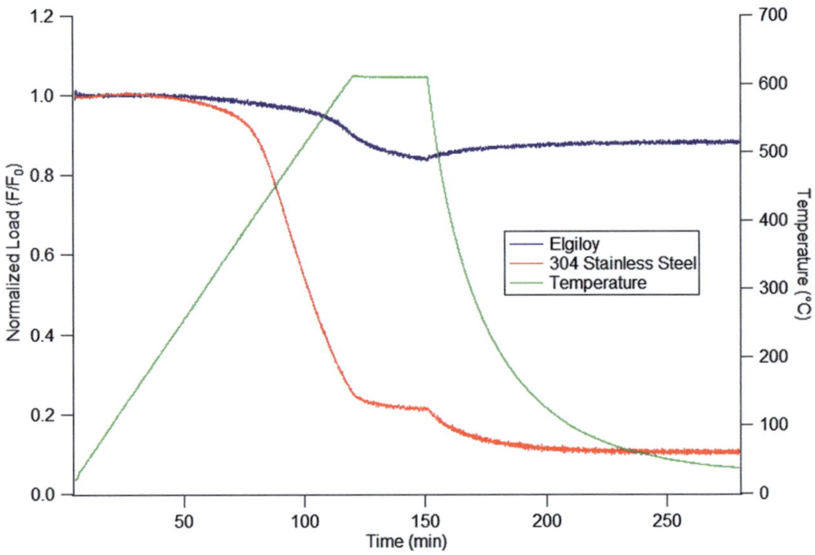

Fig. 1.13 Relaxation of Elgiloy and 304 stainless steel springs at 600 °C with initial load normalized

Also investigated in this study was the effect of heating rate. The Elgiloy spring was tested at a dwell temperature of 600 °C for 30 min at heating rates of 5, 15, and 25 °C/min (Fig. 1.14). As anticipated, the slower heating rates caused more relaxation in the spring due to the spring being held at a higher temperature for longer. Using both tests at different dwell temperatures and tests at different heating rates can combine to provide sufficient data for a creep model.

The springs themselves show the damage that can be caused by thermal creep (Fig. 1.15). For the Elgiloy springs, the temperature ramp and hold to 500 °C shows very little damage besides a slight discoloration of the spring. At 600 °C a small amount of plastic deformation can be seen as the spring has a slightly longer unloaded length. When an Elgiloy spring is tested to 700 °C however, it is significantly longer and shows a large amount of discoloration. The coil has deformed enough that gaps have formed between the loops.

The 304 stainless steel springs show a similar trend to that of the Elgiloy springs, only at lower temperatures. At 500 °C, the spring shows discoloration and some permanent extension of its length. When the springs are tested at 600 °C however, significant permanent deformation can be seen. Large gaps have formed between the loops of the spring and it appears as if the coil has turned slightly.

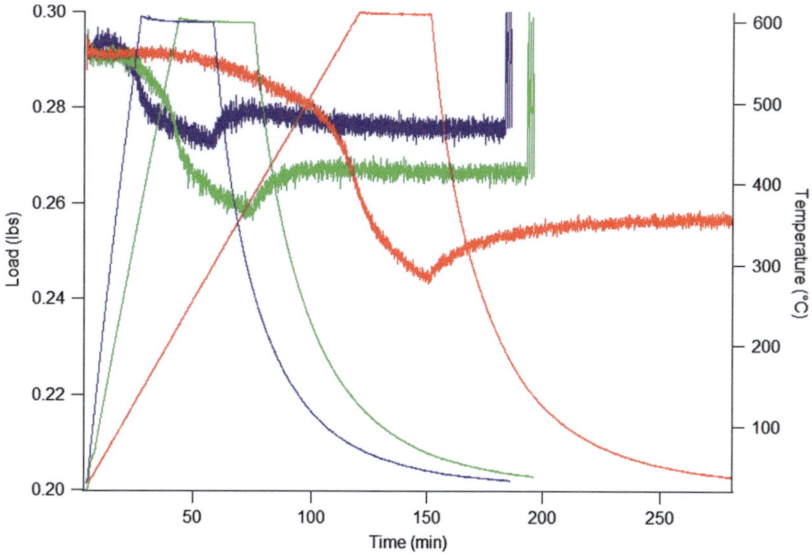

Fig. 1.14 Effect of heating rate on spring relaxation at 5, 15, and 25 °C/min

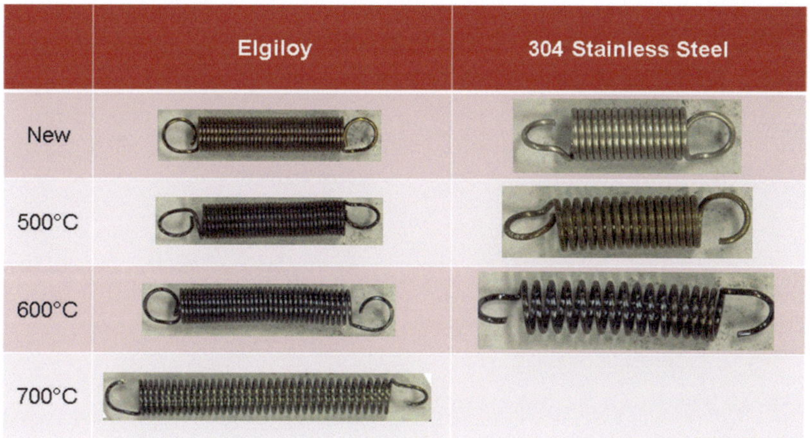

Fig. 1.15 Photos of springs after relaxation testing

1.4 Summary

Extension springs are used in mechanisms and components across a wide range of applications. They are not typically designed to sustain periods of extreme heat but it is valuable to characterize the risk of failure in abnormal thermal events. At high temperature there are a number of different modes of deformation and damage that contribute to load relaxation and thermal degradation. In this study two different material types of springs were tested at various temperatures and heating rates to help provide information to develop models for their high temperature behavior. It was found that Elgiloy performs much better than 304 stainless steel under abnormal thermal conditions.

Acknowledgements Sandia National Laboratories is a multi-program laboratory managed and operated by Sandia Corporation, a wholly owned subsidiary of Lockheed Martin Corporation, for the U.S. Department of Energy's National Nuclear Security Administration under contract DE-AC04-94AL85000.

Reference

1. R.C. Dykhuizen, C.V. Robino, Load relaxation of helical extension springs in transient thermal environments. Journal of Materials Engineering and Performance **13**(2), 151–157 (2004)

Chapter 2
Effect of Applied Temperature and Strain Rate on Laser Welded Stainless Steel Structures

Bonnie R. Antoun and Kevin J. Connelly

Abstract Sealed containers that hold organic substances can fail if organic material decomposition that occurs at elevated temperatures causes high enough pressures to cause a breach anywhere within the container or at welded or joined sections of the container. In this study, the response of stainless steel structures sealed by laser welding was of interest. Cylindrical can structures were constructed of two base materials, 304L stainless steel in tube and bar form, and joined by partial penetration laser welding. The base and weld materials contributed to the overall elastic–plastic response that led to failure in the weld region. The response of specimens constructed from sections of the cylindrical can structures was measured experimentally under thermomechanical loadings that investigated applied strain rate and temperatures (25–800 °C). Prior to testing, extensive measurements of the partial penetration weld geometry and cross section were completed on each specimen to enable correlation with measured response and failure. The experimental results of these sub-structure specimens tested at elevated temperatures are presented. Additionally, the material characterization results of the two 304L stainless steel materials used in constructing the cylindrical cans are presented.

Keywords 304L • Laser weld • Failure • Elevated temperature • Pressurization

2.1 Material

The laser welded can-like structures were constructed of two distinct forms of 304L stainless steel. The lid and base plate, shown in far right and far left, respectively, of the photograph in Fig. 2.1 were machined from 88.9 mm (3.5 in.) diameter solid bar stock. The walls of the can structures, shown third from left in Fig. 2.1, were machined from 88.9 mm (3.5 in.) outside diameter by 4.76 mm (3/16 in.) wall thickness seamless tubing. 304L stainless steel is sensitive to manufacture, form and source, so both materials were characterized individually. The foam material shown second from left in Fig. 2.1 was not characterized as a part of this project.

2.2 Tensile Specimen Geometry

Cylindrical threaded tensile specimens were removed from the two 304L materials, longitudinally along the length of the stock. For the can lid material (bar stock), the gage section was 8.89 mm (0.350 in.) diameter by 38.1 mm (1.5 in.) long and the nominal section that strain was measured over using a contacting extensometer was 20.3 mm (0.8 in.). A few specimens were also taken in the radial orientation for the lid material, these specimens were smaller with a 4.7 mm (0.185 in.) diameter gage section. For the can wall material (tube stock), the gage section was 2.54 mm (0.10 in.) in diameter and 21.6 mm (0.85 in.) in length, strain was measured over 15.75 mm (0.62 in.) of that length.

B.R. Antoun (✉) • K.J. Connelly
Sandia National Laboratories, Livermore, CA 94551, USA
e-mail: brantou@sandia.gov

B. Antoun (ed.), *Challenges in Mechanics of Time Dependent Materials, Volume 2*,
Conference Proceedings of the Society for Experimental Mechanics Series, DOI 10.1007/978-3-319-22443-5_2

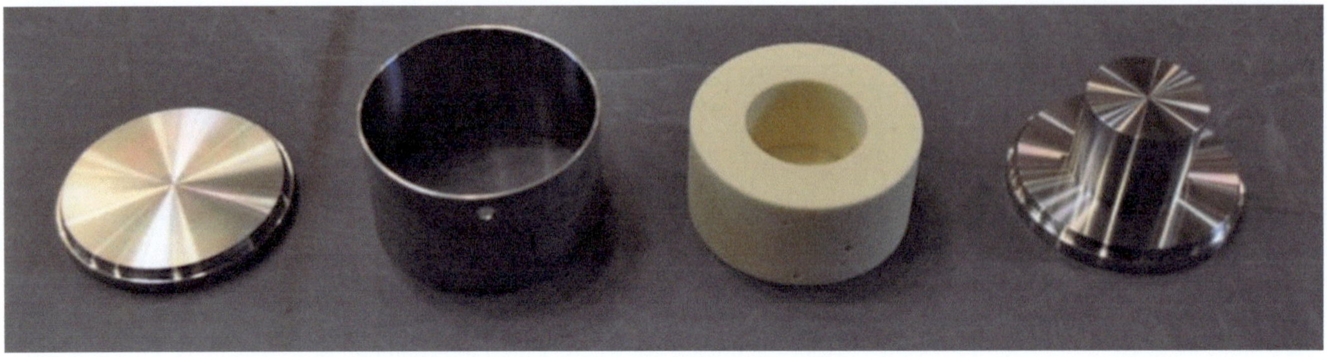

Fig. 2.1 Laser welded can structure geometry before assembling [1]

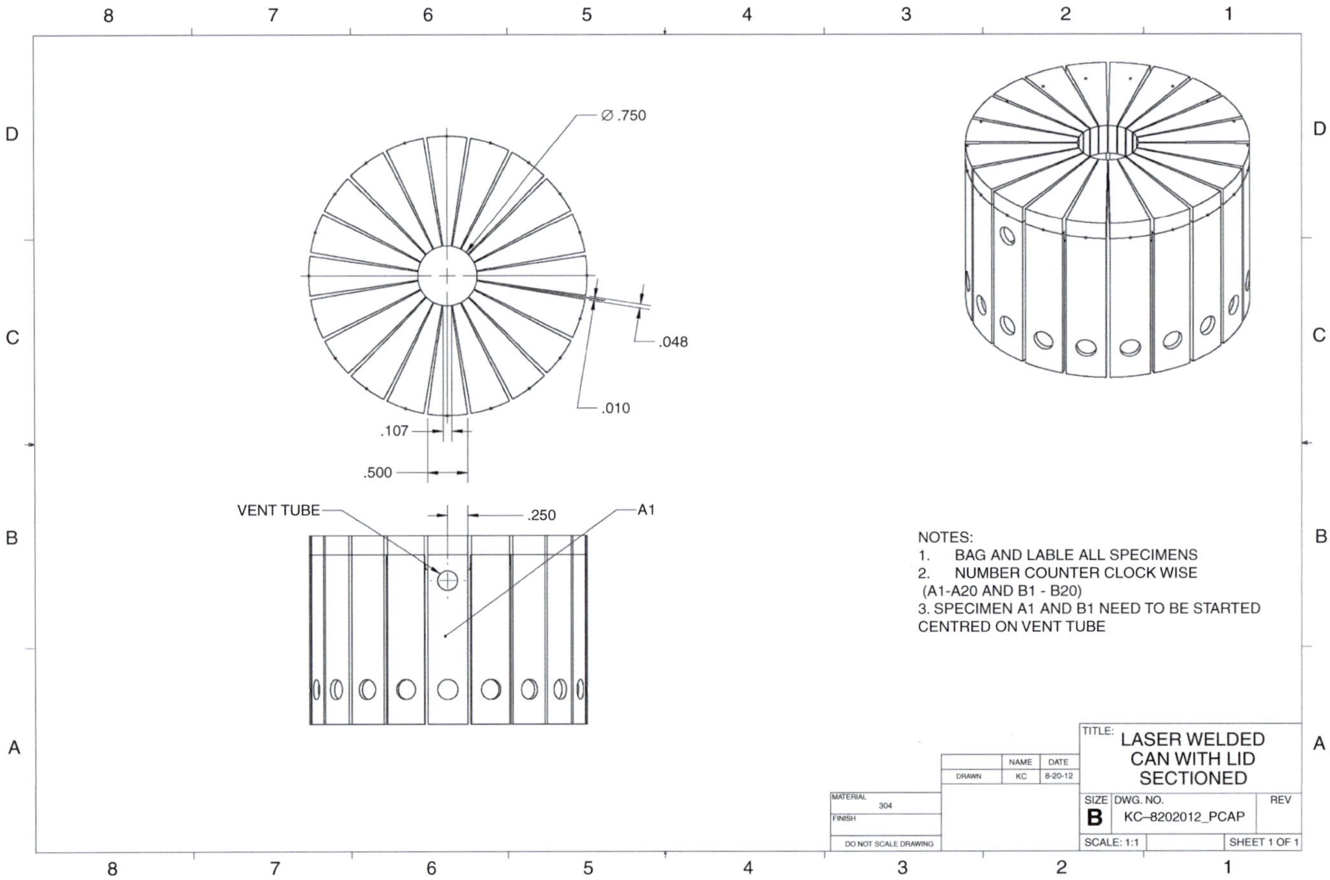

Fig. 2.2 Laser welded can sectioned into 20 wedge shaped specimens

2.3 Wedge Specimen Geometry

Wedge specimens were removed from two can structures, referred to as CAN A and CAN B. Note that these cans only had their lid welded on to the side wall portion of the structure, the base plate was not welded on since the intent was to test one welded section only during experiments and to have a means to grip and load the welded section. Figure 2.2 shows the layout for sectioning both cans into 20 wedge specimens, for a total of 40. Each wedge specimen was 12.7 mm (0.50 in.) along the arc, and labeling was completed sequentially from 1 to 20 counterclockwise from the vent tube location. Photographs of a typical wedge specimen are shown in Fig. 2.3.

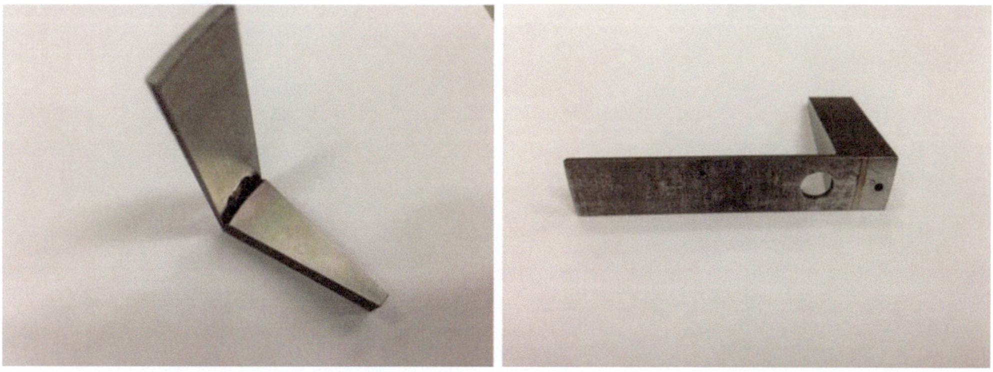

Fig. 2.3 Photographs of wedge specimens after removal from can structure

2.4 Weld Penetration Depth Quantification

In addition to tracking location relative to the vent tube, which is related to the starting and ending point of the laser welding process, the weld penetration depth was measured on both sides of each specimen. The target weld depth was 7.6 mm (0.03 in.) [1], by carefully polishing and making measurements with a measuring microscope, the weld penetration depth was quantified at 40 points around the circumference of CAN A and CAN B. The polishing and measurements were completed prior to testing. The polishing procedure has no effect on the mechanical test results.

An example of the weld cross-section measurements made for each specimen edge is shown in Fig. 2.4. Several important features, such as weld penetration, lid-to-wall clearance, geometry of the weld cross-section, and void size and depth were captured. A summary of the weld penetration measurements is shown in Figs. 2.5 and 2.6 for CAN A and CAN B. Note that the actual weld penetration depth is not constant and actually varies along the circumference. The weld penetration depth varies from less than 0.635 mm (0.025 in.) to just over 0.762 mm (0.030 in.). Penetration was noticeably deeper for CAN A than CAN B, based only upon these discrete measurement points.

2.5 Experimental Results: Tensile Characterization

Details of the experimental methods used for the elevated temperature tensile tests can be found elsewhere [2]. The nominal strain rate was 0.001/s; limited additional tests were conducted at other strain rates at a few temperatures [3]. Tensile tests were conducted at several temperatures: 25, 100, 200, 400, 600, 700 and 800 °C. Most specimens were in the axial orientation, but some tests were conducted on the lid material at room temperature, 600 and 800 °C in the radial orientation to get an idea of orientation dependence. It was not possible to fit tensile specimens in the radial orientation for the wall (tubular) material.

The results for the can lid material in the axial direction are shown in Fig. 2.7. The effect of annealing the specimens was marked, especially at the lower temperatures. As the test temperature increased, this effect became less noticeable and by 700 °C, fairly indistinguishable. In the radial orientation, a larger amount of scatter in the failure strain was noted. The hardening behavior was also found to be somewhat different than the axial orientation, with failure occurring just after necking [3]. The results for the can wall material are shown in Fig. 2.8. The effect of annealing the specimens was even more pronounced that for the lid (bar stock) material, even though the tube stock material was reportedly already annealed. As the test temperature increased, this effect became less noticeable and by 700 °C, or certainly 800 °C, fairly indistinguishable.

2.6 Experimental Results: Wedge Specimens Loaded in Bending/Shear and Tension

Custom fixtures were designed to test the laser weld response in the wedge specimens up to 800 °C under two types of loading: tension and combined bending/shear. Details of the fixture designs, test methods and measurements can be found in [3]. The effect of loading rate was studied using two rates. In bending/shear, a few experiments were conducted at 25 and 800 °C, the response was very high amount of deflection without failure. This deformation was found to be unrepresentative of the

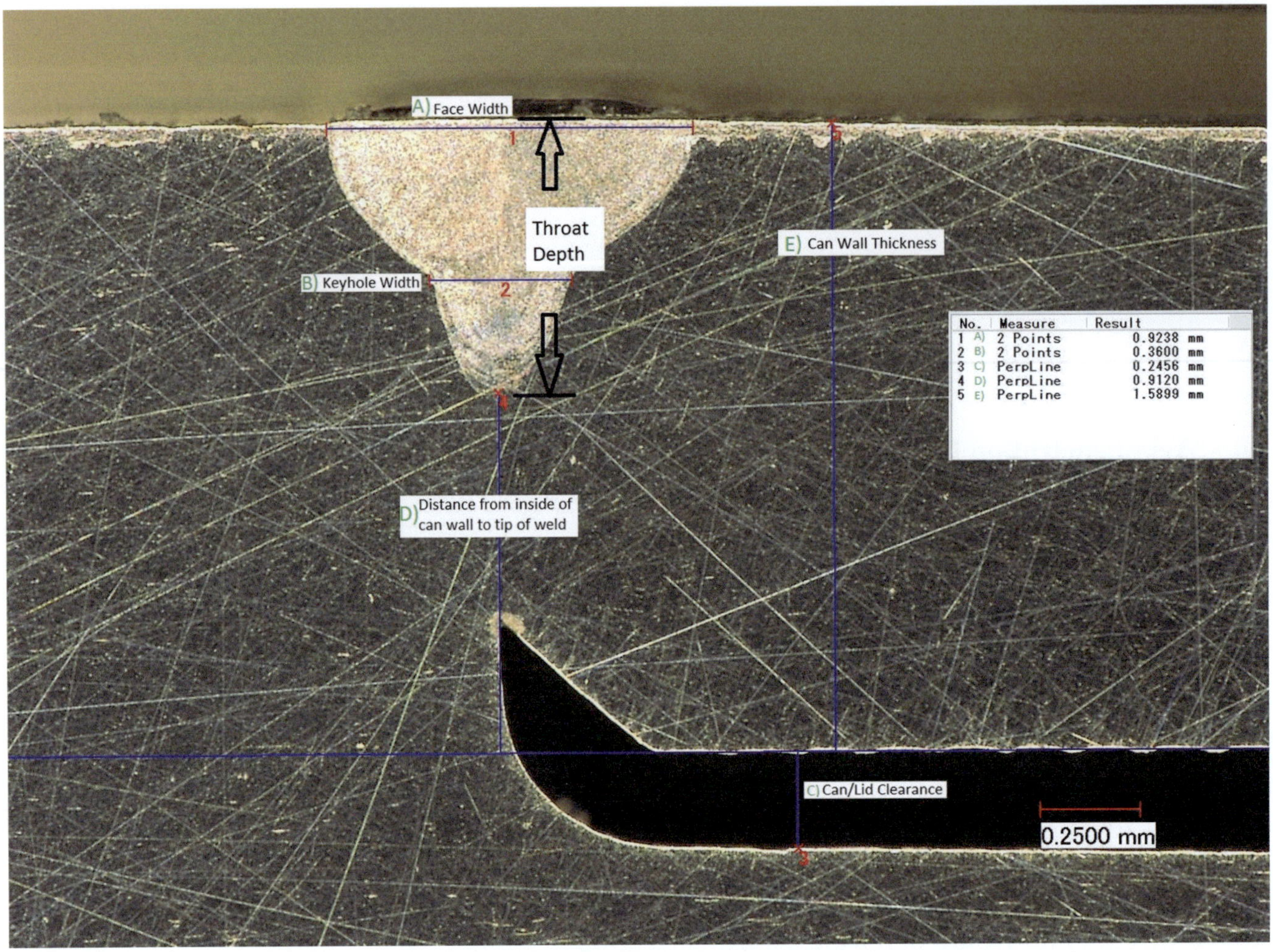

Fig. 2.4 Measurements of laser weld used to determine weld penetration depth and geometry details

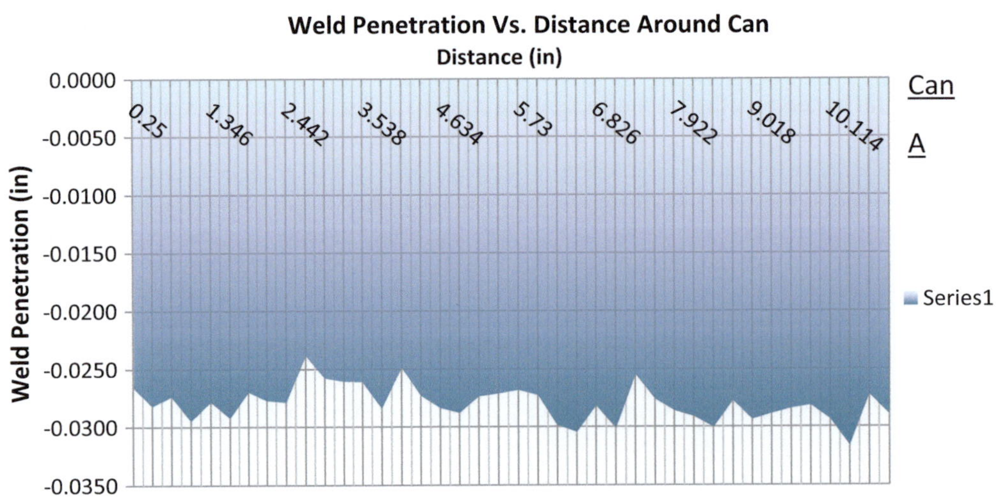

Fig. 2.5 Measured weld penetration along weld line of CAN A

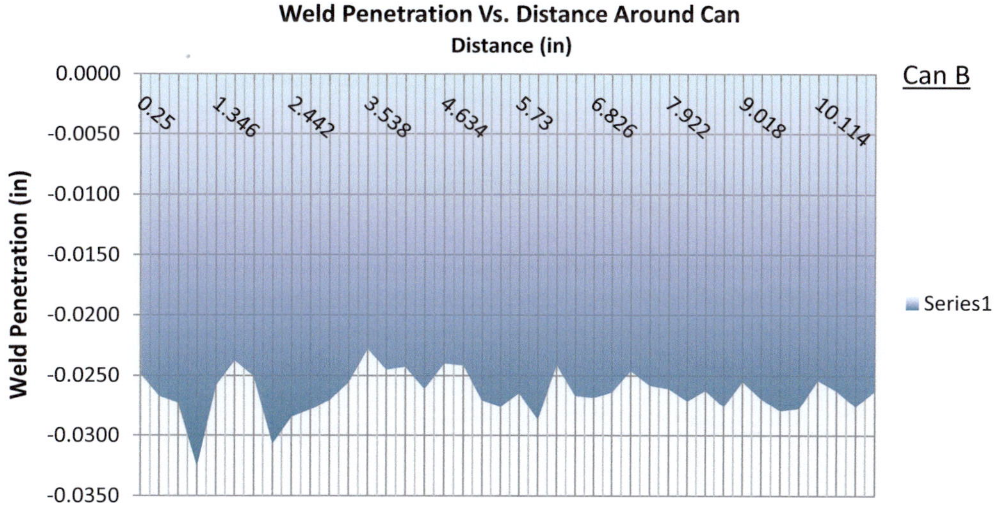

Fig. 2.6 Measured weld penetration along weld line of CAN B

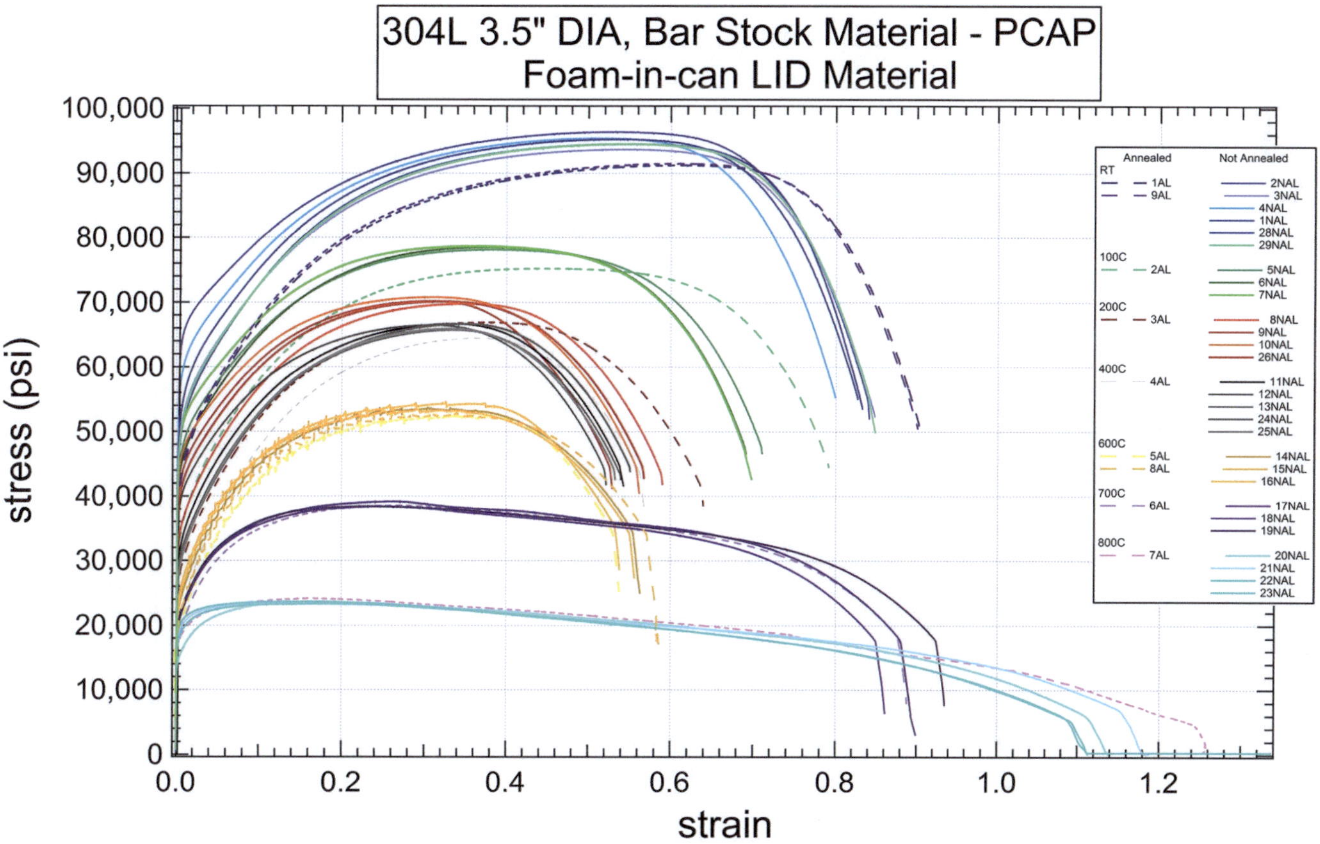

Fig. 2.7 Engineering stress versus engineering strain tensile curves for can lid 304L material

failure of a whole can due to differences in boundary conditions. Wedge specimens tested in tension were successful however; tests were conducted at two loading rates at temperatures of 25, 400, 600, 700 and 800 °C.

The results are shown collectively in the graph in Fig. 2.9. Note that the darker lines are always the faster of the two loading rates. At 25 °C, the faster rate results in a slightly higher load response, as would be expected for a typical strain rate effect. At 400 °C, there is some dynamic strain aging evident, which causes the load drops and recoveries. The load levels are approximately the same for the two loading rates, but the total displacement is lower for the higher strain rate. The dif-

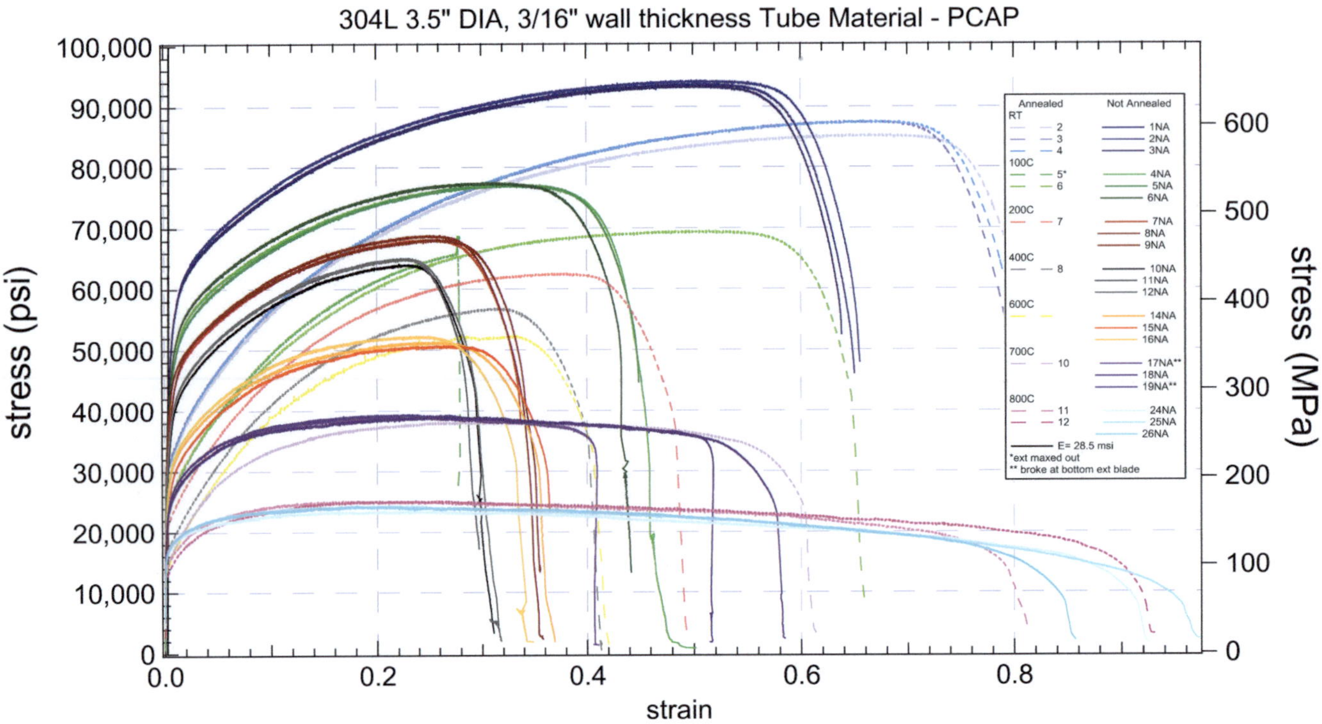

Fig. 2.8 Engineering stress versus engineering strain tensile curves for can wall 304L material

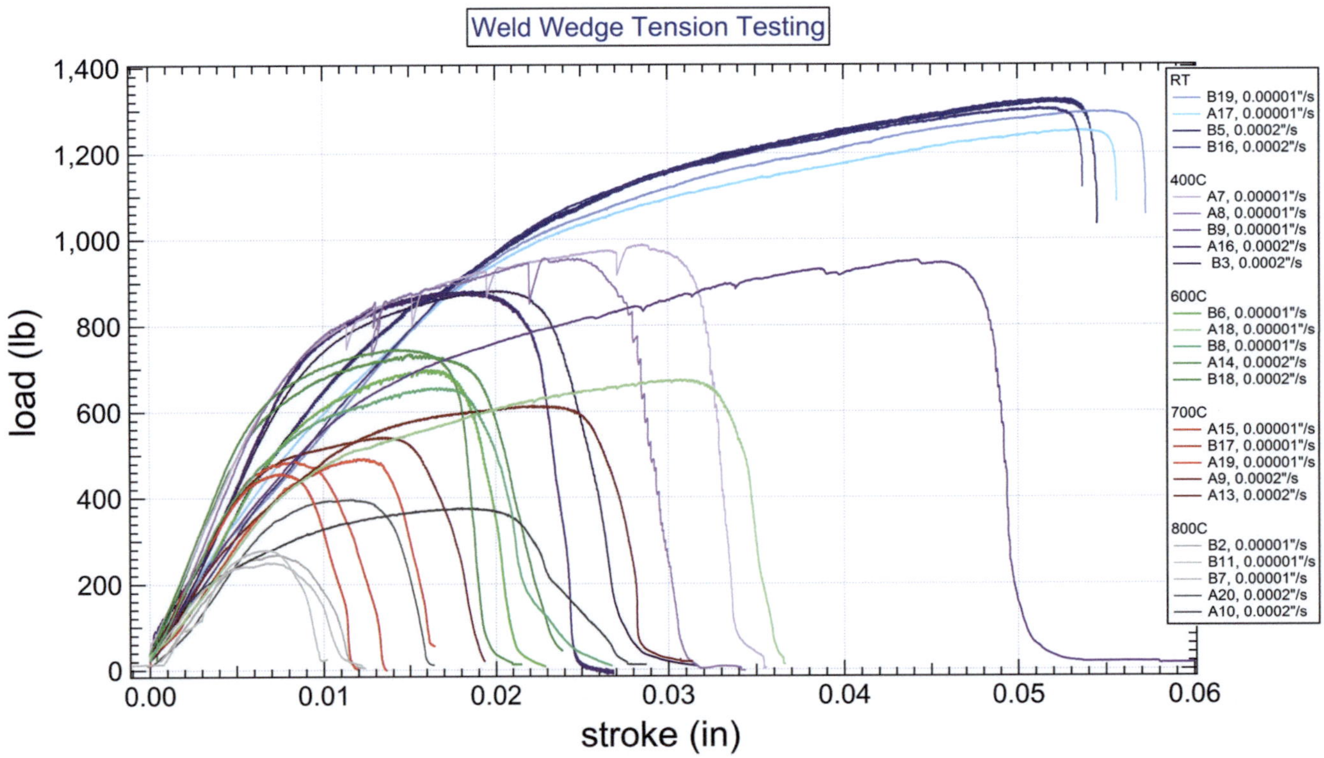

Fig. 2.9 Response of laser welded wedge tension tests at several temperatures and two loading rates

ference in total displacement is neither unexpected nor expected, but the load levels achieving the same level implies less rate dependence at 400 °C than at RT. This actually is expected, based on similar results in the base 304L stainless steel materials (lid and wall) that have been joined by laser welding. At 600 °C, the response is back to what one would expect with normal rate dependence and this continues for the 700 and 800 °C experiments. See [4] for similar findings of rate effects in 304L stainless steels.

2.7 Summary

The effect of loading rate and temperature on the response of laser welded can structures has been quantified using specially designed specimens and fixtures. This data, along with tensile characterization results of the can bulk materials joined by laser welding, is being used to assist with modeling of the whole can structures that are subjected to extreme temperatures.

Acknowledgements Sandia National Laboratories is a multi-program laboratory managed and operated by Sandia Corporation, a wholly owned subsidiary of Lockheed Martin Corporation, for the U.S. Department of Energy's National Nuclear Security Administration under contract DE-AC04-94AL85000.

References

1. J.M. Suo-Anttila, J. Koenig, A. Dodd, C. Robino, E. Quintana, *Thermal Mechanical Failure Exclusion Region Barrier Breach Foam Experiments, SAND2013-7346* (Sandia National Laboratories, Albuquerque, 2013)
2. J.F. Dempsey, B.R. Antoun, V.J. Romero, G.W. Wellman, W.M. Scherzinger, S. Grange, Temperature dependent ductile material failure constitutive modeling with validation experiments, in Proceedings of the SEM 2012 Annual Conference and Exposition on Experimental and Applied Mechanics, Costa Mesa, 11–14 June 2012
3. B.R. Antoun, K. Connelly, R.R. Barth, *Measured Response of Laser Welded 304L Stainless Steel Cylindrical Cans to Abnormal Thermal-Structural Loadings, SAND 2015-XXXX*, (2015, in press)
4. B.R. Antoun, R.S. Chambers, J.M. Emery, R. Tandon, Small strain plasticity behavior of 304L stainless steel in glass-to-metal seal applications, in Proceedings of the SEM 2014 Annual Conference and Exposition on Experimental and Applied Mechanics, Greenville, 2–4 June 2014

Chapter 3
Process Modeling and Experiments for Forging and Welding

Arthur A. Brown, Lisa A. Deibler, Lauren L. Beghini, Timothy D. Kostka, and Bonnie R. Antoun

Abstract We are developing the capability to track material changes through numerous possible steps of the manufacturing process, such as forging, machining, and welding. In this work, experimental and modeling results are presented for a multiple-step process in which an ingot of stainless steel 304L is forged at high temperature, then machined into a thin slice, and finally subjected to an autogenous GTA weld. The predictions of temperature, yield stress, and recrystallized volume fraction are compared to experimental results.

Keywords Modeling • Forging • Welding • Microstructure • Recrystallization

3.1 Introduction

The microstructure and properties in mechanical components are highly sensitive to the processing conditions under which the component is manufactured. The ability to predict the evolution of material properties during thermal-mechanical processing can be valuable in component design. Process conditions can be optimized to achieve desirable yield stress, microstructure, and residual stress. To this end, we are developing the capability to track material changes through numerous possible steps of the manufacturing process, such as forging, machining, and welding. In this work, experimental and modeling results are presented for a multiple-step process in which an ingot of stainless steel 304L is forged at high temperature, then machined into a thin slice, and finally subjected to an autogenous gas tungsten arc (GTA) weld. The predictions of temperature, yield stress, and recrystallized volume fraction are compared to experimental results.

3.2 Material Model

A treatment of the kinematics and thermodynamics of the material model utilized in the simulations presented here is documented in [1]. The internal state variable model accounts for rate and temperature dependence, and tracks the evolution of static and dynamic recrystallization. The model was validated in [2], where it was demonstrated to accurately predict the final yield strength at multiple locations in forgings of different geometries and forging temperatures. In this work, the same material parameter set is used as in [2].

3.3 Experiments and Simulations

3.3.1 Forging

A cup-shaped forging from 304L stainless steel was produced using four forging stages, where the final stage was performed at 829 °C (1525 °F) and was preceded by a furnace anneal at 982 °C (1800 °F) and a flattening operation at 816 °C (1500 °F). For the purposes of modeling, it was assumed that the annealing removed the effects of the previous stages, so

A.A. Brown (✉) • L.A. Deibler • L.L. Beghini • T.D. Kostka • B.R. Antoun
Sandia National Laboratories, Livermore, CA 94551, USA
e-mail: aabrown@sandia.gov

B. Antoun (ed.), *Challenges in Mechanics of Time Dependent Materials, Volume 2*,
Conference Proceedings of the Society for Experimental Mechanics Series, DOI 10.1007/978-3-319-22443-5_3

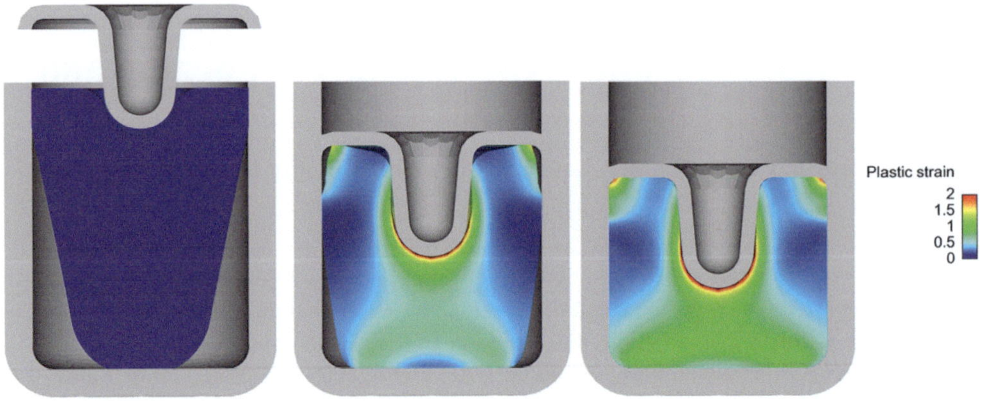

Fig. 3.1 The simulated evolution of plastic strain during the flattening operation and the final forging stage

only the flattening and the final forging stage were simulated (see Fig. 3.1). The thermal and mechanical codes used are part of Sandia's Sierra Multiphysics Code (see [3] and [4]). The simulations include the effects of furnace heating, die chill, radiation, convection, and heat generation due to plastic dissipation. It is assumed that 95 % of the plastic work is dissipated as heat.

In order to obtain microstructures for as-forged and post welded that could be reasonably compared, the forging was cut in half along the vertical axis, then quarter inch slices were cut from each half. The U-shaped slices were then cut along the mirror axis of the U (Fig. 3.2) and two of the mating surfaces were used to investigate the changes in microstructure and hardness in the heat affected zone.

3.3.2 *Welding*

An autogenous gas tungsten arc weld was performed on one side of one of the forging slices. Eight thermocouples were spot-welded in various locations on both the front and back surface of the forging slice (Fig. 3.2). The slice was supported by a thin metal frame to raise the sample off of the backing plate and provide room for the thermocouples on the back of the part (Fig. 3.3). Tabs were spot welded on to the top and bottom of the slice to avoid the effects of starting and stopping the weld on the heat affected zone in the part. The whole assembly was clamped into a fixture, and a GTA weld was performed at 130 A, travel speed 2.5 in. per minute.

For the modeling of the welding process, the tabs and the clamping shims were included. The stress state and internal state variables were mapped from the forged mesh to the mesh used for the welding simulation using Exomerge [5]. Figure 3.4 shows the temperature contour in the specimen and tabs during the simulation. The heat flux input to the top surfaces of the tabs and forged specimen was optimized to get the simulation temperature histories at the thermocouple locations to match the experimentally measured histories (Fig. 3.4). Only data from the first five thermocouples was obtained, and the experimental output from thermocouple 4 is not accurate after the peak temperature since the torch knocked off the thermocouple shielding as it passed by. A fairly good representation of the thermal histories is obtained through the simulation of the welding process.

3.3.3 *Measurements of Recrystallized Volume Fraction and Hardness*

Both the welded and the un-welded pieces were sectioned for metallography, as shown in Fig. 3.2. Samples were polished to a 0.3 μm finish. Hardness testing was performed on polished samples. Vickers hardness testing was performed with 100 g load, and 15 s dwell time, keeping track of the X and Y position of each indent. In order to compare with the modeling results, the Vickers hardness needed to be converted to yield strength. A conversion between Knoop hardness and yield strength was available from [6]:

$$\sigma_y = 2.673 \cdot 10^6 HK - 1.489 \cdot 10^8 \text{ Pa} \tag{3.1}$$

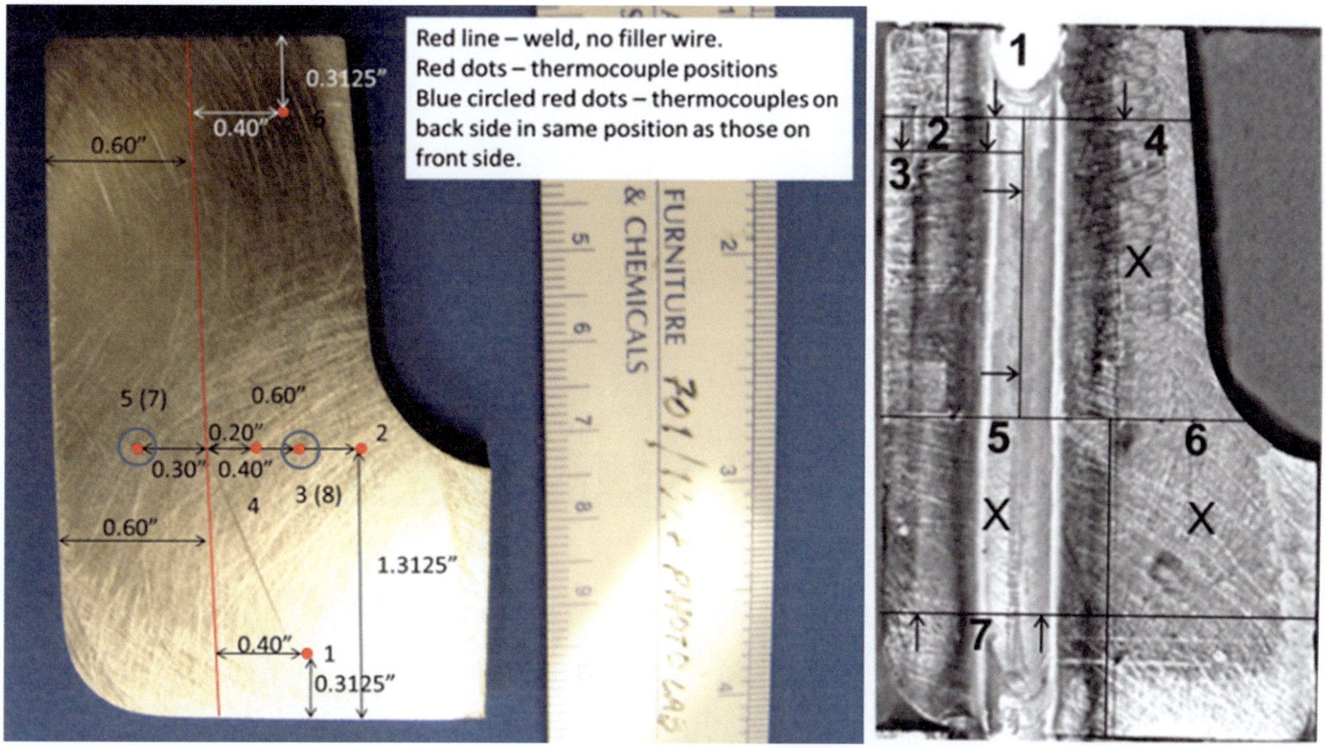

Fig. 3.2 The slice before welding with locations of the thermocouples indicated (*left image*). A map of the metallographic sections on the welded part is shown on the right. Planar sections are indicated by X, transverse or longitudinal sections are arrowed to indicate the face mounted. The as-forged part was cut similarly

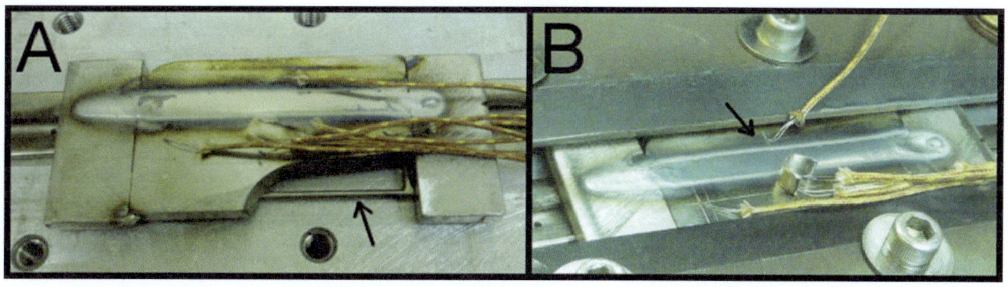

Fig. 3.3 The welded part, showing the start and stop tabs at each end of the weld and the support framework (*left image*). The part was clamped during welding (*right image*)

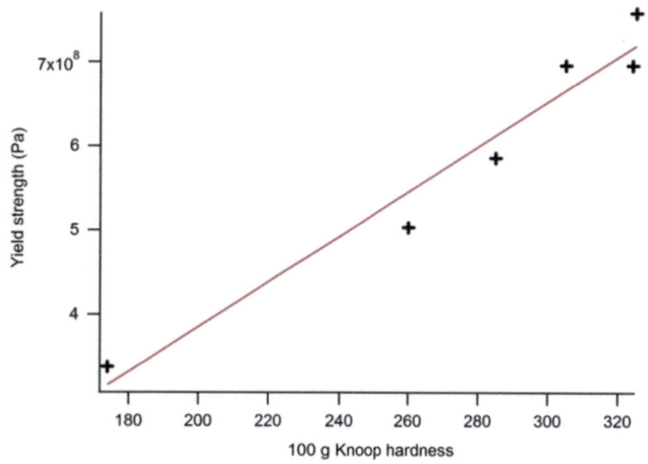

Fig. 3.4 The temperature profiles from the thermocouples. The simulation temperatures at the same locations are shown for comparison

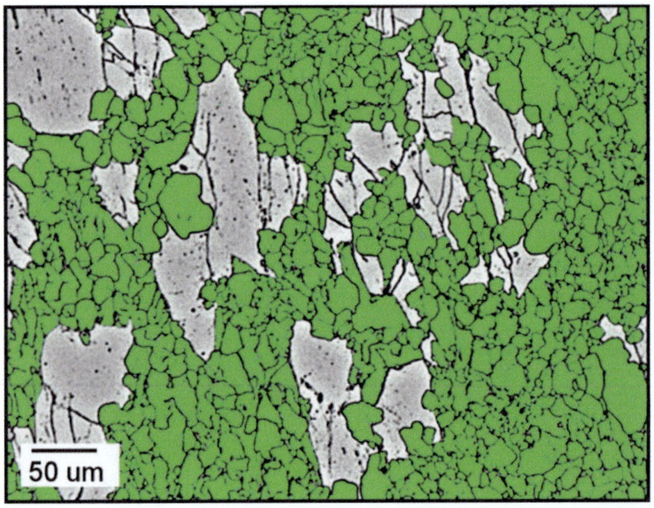

Fig. 3.5 Data to establish conversion between Knoop hardness and yield strength

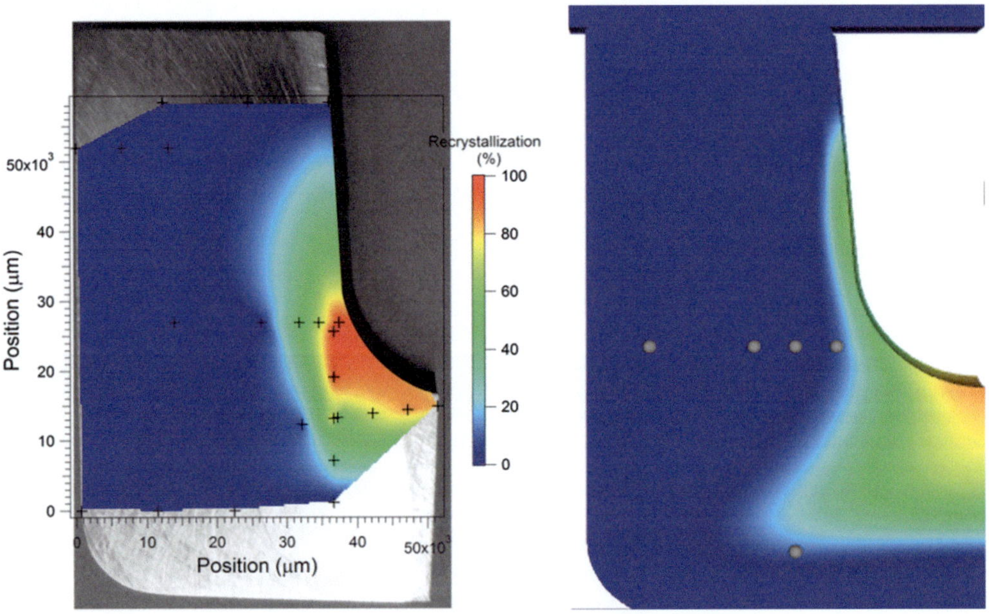

Fig. 3.6 An example of the microstructure imaging used to determine the recrystallized volume fraction. This image came from the as-forged condition (bottom-right of section 6), with measured values of 48 % recrystallization and 233 Vickers hardness

The following conversion between Vickers and Knoop hardness was obtained by performing ten measurements of each type on forging specimens subjected to various heat treatments to achieve various hardness values:

$$HK = 1.115HV - 9.801 \tag{3.2}$$

The above two equations were used to convert the Vickers hardness measurements to yield strength (Fig. 3.5). The yield strength derived from this calculation is plotted in Figs. 3.8 and 3.10 for the as-forged and as-welded specimens. The markers indicate where data was collected.

After hardness testing was complete, samples were etched for microstructural imaging with 60 % HNO_3, 40 % DI H_20 electrolytic at 1.5 V for 20–60 s (ASTM E407-219). Optical micrographs were taken at regular intervals on both sets of samples, keeping track of the position of each micrograph. At each location where recrystallization data was desired, three separate fields of view were captured to allow for statistical analysis of data. The amount of recrystallization was determined by visually defining which grains appeared recrystallized (generally smaller with a rounder outline) and measuring the volume fraction of recrystallized material using Clemex image analysis tool (see Fig. 3.6).

3.4 Results

The results of the recrystallized volume fraction measurements for the forged specimen before the welding operation are plotted in Fig. 3.7, and the results corresponding to the specimen after welding are shown in Fig. 3.9. Individual data points are marked in the figure; the color map is an extrapolation between data points. The average standard deviation of the non-zero or 100 % recrystallization measurements is ±3 %. Because of the size of samples for metallographic sectioning, the entire area of the forging could not be evenly sampled. The model predictions of the recrystallized volume fraction are also shown in Figs. 3.7 and 3.9. The simulations predict that the forging process alone leads to almost complete recrystallization right under the punch. The general trends are consistent, in that both the experiments and the simulations show significant recrystallization on the sidewalls and bottom of the inner forging surface, where the punch produced the highest plastic strains in the forging (Fig. 3.1).

The model prediction of the as-forged yield stress is shown in Fig. 3.8. The highest strength is at the bottom of the specimen. This is due to die chill; the bottom is cooled due to conduction to the die, and thus is plastically deformed at a lower temperature, which leads to more stored energy in the dislocation structure. The strength is predicted to be low under the punch, due to the recrystallization that wipes away the dislocation structure there, softening the material. This effect is not evident in the experimental results. It is not clear to the authors why the measured yield stress is high in some locations that exhibited significant recrystallization. This will be further investigated in the future through additional measurements. There are obviously some artificial artifacts due to the interpolation between experimental measurements that is used to create the experimental contour plots. (In fact, Fig. 3.10 shows that the experimental results in the as-welded sample do exhibit low yield stresses under the punch, which we do not believe is due to the weld).

The autogenous weld that travels up the specimen sidewall introduces high local temperatures, which leads to melting and solidification, as well as a heat-affected zone that experiences recrystallization and softening (Figs. 3.9 and 3.10). The model in its current form does not account for melting and solidification, but as can be seen in the figures, does predict the recrystallization and local softening in the weld region and heat-affected zone.

Uncertainty quantification will be performed as future work. The predictions of room temperature yield stress and recrystallized volume fraction are very sensitive to certain simulation inputs, such as the time it takes to get the forging out of the die and into the quench bath. The results presented here are for nominal input values, but future work will be to include the effects of uncertainties on the final predictions.

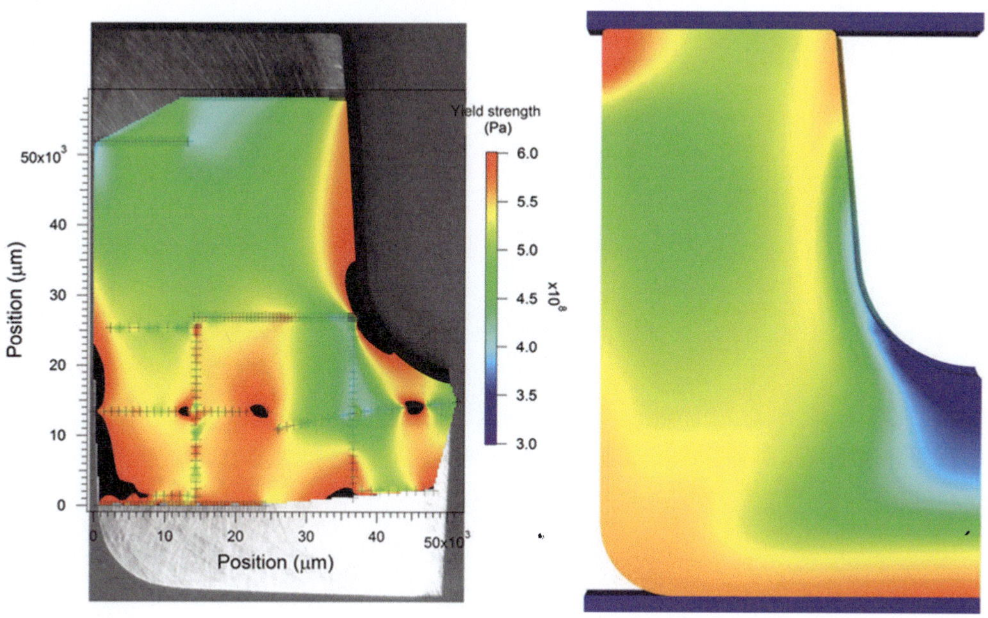

Fig. 3.7 The measured (*left*) and predicted (*right*) recrystallized volume fraction before the welding operation

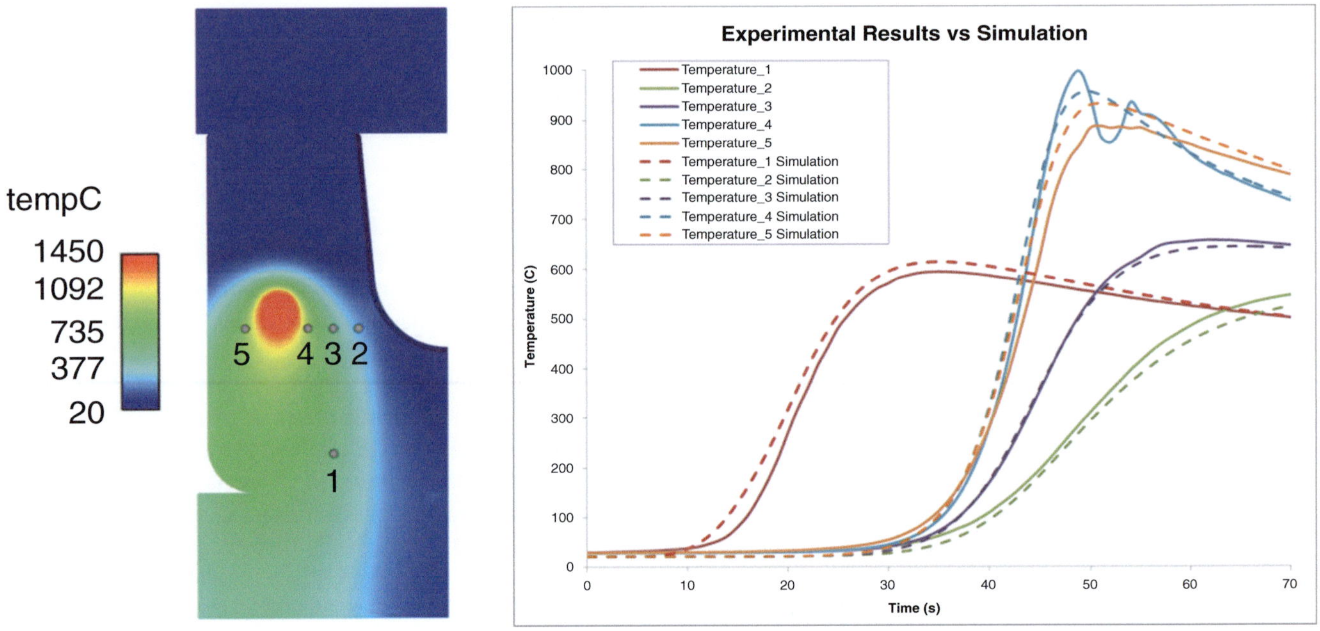

Fig. 3.8 The measured (*left*) and predicted (*right*) yield strength of the forging before the welding operation

Fig. 3.9 The measured (*left*) and predicted (*right*) recrystallized volume fraction after the welding operation

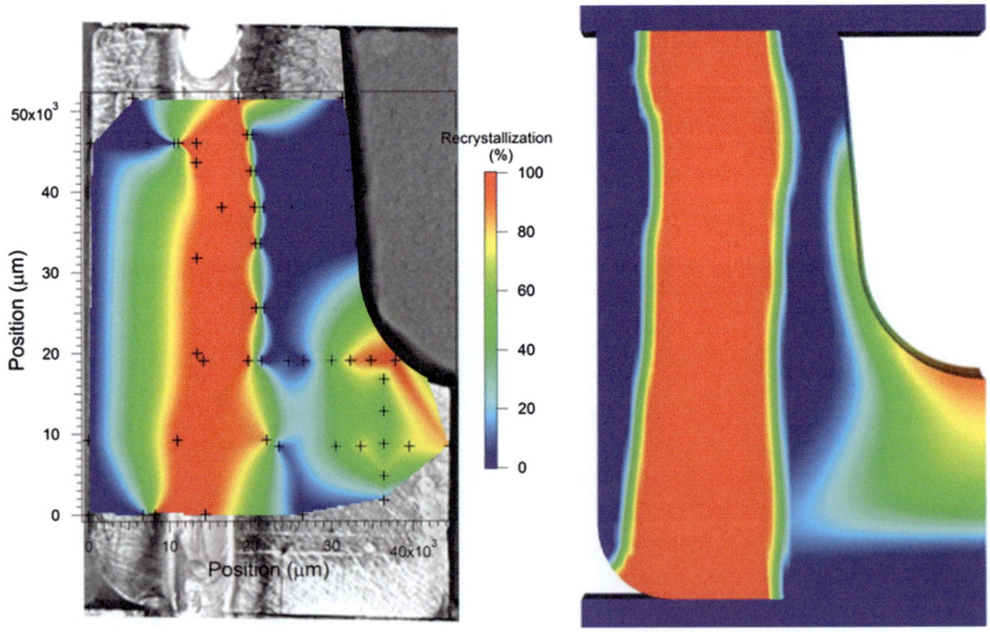

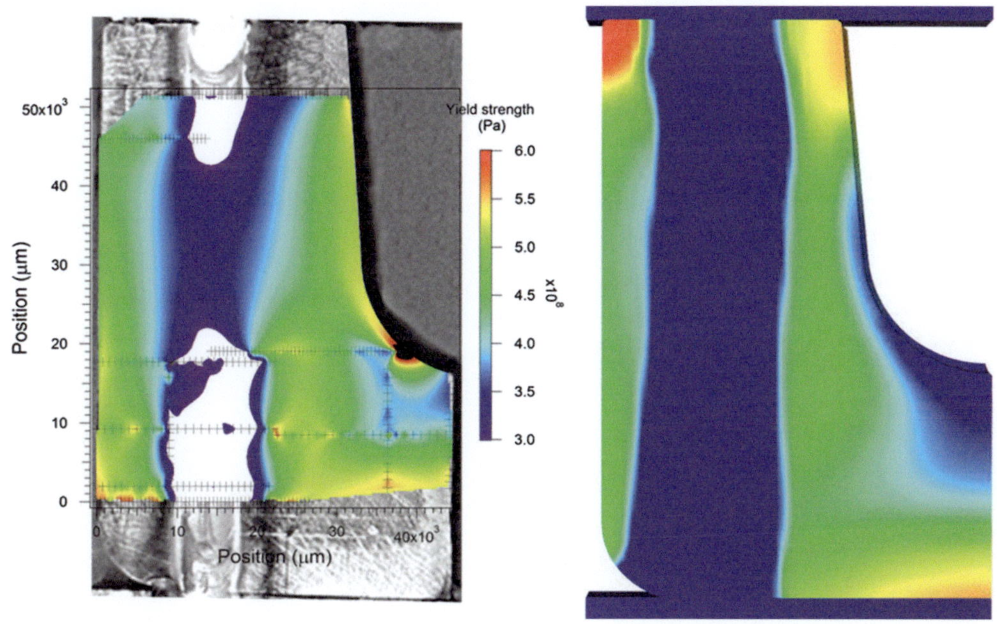

Fig. 3.10 The measured (*left*) and predicted (*right*) yield strength of the forging after the welding operation

3.5 Conclusions

This paper demonstrates the capability to model multiple stages of manufacturing processes such as forging and welding. There are some discrepancies between predictions and experimental measurements that still need to be addressed, but the general trends of recrystallization and softening are captured in the as-forged and as-welded specimens.

Acknowledgements This work was performed at Sandia National Laboratories. Sandia is a multiprogram laboratory operated by Sandia Corporation, a Lockheed Martin Company, for the United States Department of Energy under contract DEAC04-94AL85000.

References

1. A.A. Brown, D.J. Bammann, Validation of a model for static and dynamic recrystallization in metals. Int. J. Plast. **32–33**, 17–35 (2012)
2. A.A. Brown, T.D. Kostka, B.R. Antoun, M.L. Chiesa, D.J. Bammann, S.A. Pitts, S.B. Margolis, D. O'Connor, N.Y.C. Yang, Validation of thermal mechanical modeling for stainless steel forgings. Comput. Plast. XI: Fundam. Appl. 1153–1164, (2011)
3. SIERRA Solid Mechanics Team. Adagio 4.14 User's Guide. Sandia Report 2009-7410, (2009)
4. P.K. Notz, S.R. Subia, M.M. Hopkins, H.K. Moffat, D.R. Noble, Aria 1.5 User Manual. Sandia Report 2007-2734, (2007)
5. T.D. Kostka, Exomerge User's Manual: A lightweight Python interface for manipulating Exodus files. SAND2013-0725, (2013)
6. S. Robinson, C. Story, Unpublished data for converting Knoop hardness to yield stress for 304L stainless steel

Chapter 4
Time-Dependent Viscoplastic Model for Dislocation Generation During the Cooling Process in the Silicon Ingot

Maohua Lin, Qingde Chen, and C.T. Tsai

Abstract Silicon growth is a process in which a silicon ingot is solidified from the melted and then cooled to the room temperature through the control of multi-heater. Dislocation densities are generated in the ingot by excessed thermal stresses caused by the nonuniform temperature field in the ingot. The generation of the dislocation density is considered as s a process of viscoplastic deformation. A three dimensional transient finite element model based on the Haasen viscoplastic constitutive model (HAS) is developed to evaluate the dislocation densities generated in silicon ingots grown by directional solidification process. The stress fields and dislocation densities generated in silicon ingots are the two major parameters for the evaluation of ingot quality. These two results calculated by HAS model are compared with those obtained from CRSS model. The result demonstrates that HAS model is more accurate than CRSS model for the calculation of dislocation densities and stresses during the cooling process of silicon ingot because of the consideration of time-dependent viscoplastic deformation in HAS model.

Keywords Dislocation density • Thermo-stress • HAS model • CRSS model • Time-dependent

4.1 Introduction

Multicrystalline silicon solar cells have occupied the mainstream PV market for many years, while directional solidification has been commonly used for producing multicrystalline silicon ingots [1]. The current main challenge on the ingot growth is to optimize growth process conditions to grow low dislocation density silicon ingots at low cost. In order to reduce the dislocation generation during the growth process, models for the relationship between the dislocation generation and crystal growth processes have to be established. Since Taylor, Orowan, and Polanyi put forward the concept of dislocation [2], it have been widely used in major areas for decades and was first proposed in the diamond structure by Haasen [3] who came up with the relationship of plastic strain rate and the dislocation proliferation rate with temperature, stress and dislocation density. The model is extended after considering the dislocation dynamics and interactions based on experimental observations by Alexander Haasen [4], Yonenaga, Sumino and Suezawa [5, 6], where yield condition was extended to the deformed state. Then they [7, 8] studied the influence of impurities on dislocation interaction during the crystal growth process by experimental means which was considered in the model Hassen-Sumino and verified by calculation by Tsai [9]. Moosbmgger [10] introduced the variable dislocation arrangement to calculate the evolution of the dislocation structure which was used in the studies of deformation and dislocation proliferation of CdTe. Pendurti [11], considering the release and interactive influence of the dislocations, took into account the multiple slip system dislocation interaction types, and established a new model from the dislocation nucleation, movement and proliferation.

Many researchers [12–19] have used the CRSS model and HAS model to calculate the stress and dislocation including the factors of actual industry process. Jordan [20, 21] studied the twelve sliding systems of the silicon and proposed CRSS model by calculating the shear stress of sliding systems and then predicted the dislocations multiplication, where the coupling between dislocation multiplication and viscoplastic deformation process was not considered. He used this model to study the growth on the CZ method of GaAs and the simulation results highlight the temperature field control was an important impact on dislocation during III–V compound semiconductor growth process. Tsai proposed the first model to successfully calculate the quantity of dislocation density generated in Si, GaAs and InP crystal grown from the melt based on the Haasen-Sumino constitutive model [9, 22–24], which considered the macroscopic plastic deformation to the microscopic dislocation density for the diamond

M. Lin • Q. Chen • C.T. Tsai (✉)
Department of Ocean and Mechanical Engineering, Florida Atlantic University, Boca Raton, FL 33431, USA
e-mail: tsaict@fau.edu

B. Antoun (ed.), *Challenges in Mechanics of Time Dependent Materials, Volume 2*,
Conference Proceedings of the Society for Experimental Mechanics Series, DOI 10.1007/978-3-319-22443-5_4

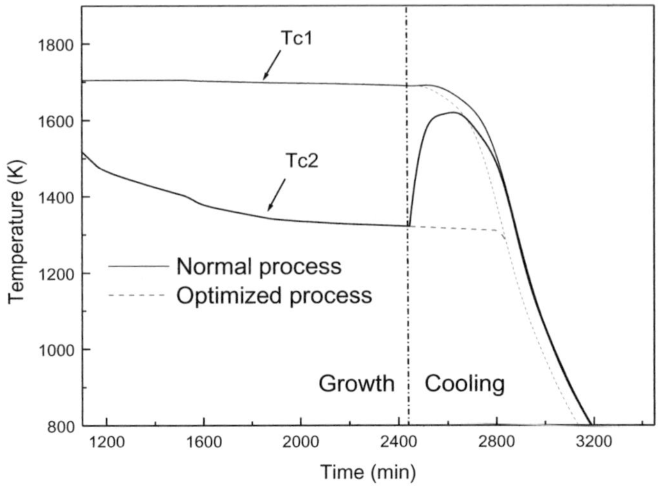

Fig. 4.1 Schematic diagram of normal and optimized processes expressed by variation of Tc1 and Tc2 with time

structure crystal [9, 23, 24]. This parameters of Hassen-Sumino model for Silicon, GaAs and InP have been identified through experimental studies by Sumino [5, 6, 25]. Then Tsai [24], considering the crystal orientation on dislocations effect, calculated the dislocation of each sliding system which was applied (001) in GaAs growth and studied the influence of the different growth direction of the dislocations [26]. Chen et al. [12, 14, 27] also employed the Hassen-Sumino model to calculate the dislocation density generated during the casting and cooling of the polycrystalline silicon crystal growth, where the silicon ingot temperature differences in a local region were the direct cause of a large amount of dislocations.

In the current direct solidification process, the ingot bottom is first rapidly heated to achieve uniform temperature to minimize thermal stresses by homogenizing the ingot temperature until it reaches the temperature on the top of ingot, holding the temperature for several tens of minutes and then slowly cooling down. However, the cracking can be avoided to some extent by using this process, but may also lead to the generation of a large number of dislocations. The stresses are actually already mostly released while it would inevitably induce new thermal stresses and hence further multiplication of dislocations. Even though many methods are made to solve the problem in the cooling process, such as tiny time adjustment based on the thermo couple trace data, the overall grasping of the process optimization for dislocation reduction hasn't been found.

In a previous study, a simplified CRSS model was used for the comprehensive analysis of dislocation multiplication in silicon ingot. The top central temperature Tc1 and the bottom central temperature Tc2 of the ingot for the normal process, shown by the solid lines shown in Fig. 4.1, have been modified to those shown in solid lines. The main change is that instead of raising Tc2 to be close to Tc1, Tc2 follows its previous slow cooling curve and then decreases with Tc1. The thermal distribution in the ingot generated by the four cooling processes shown in Fig. 4.1 have been adopted [28, 29] to calculate the stresses and dislocation densities generated in the cooling silicon ingot. In this paper, a 3-dimentional transient finite element model based on the Hassen-Sumino model is developed to predict the dislocation multiplication during the cooling process of crystal growth.

4.2 3 Dimensional Time-Dependent Haasen-Sumino Viscoplastic Model

In the Haasen-Sumino model [3, 4], the viscoplastic strain components *εijc* and dislocation density multiplication can be given by:

$$\varepsilon_{ij}^{c} = \int_{0}^{t} \dot{\varepsilon}_{ij}^{c}\, dt \tag{4.1}$$

$$\dot{\varepsilon}_{ij}^{c} = \frac{1}{2} b * k_0 * N_m * \exp(-Q/kT)\left(\sqrt{J_2} - A\sqrt{N_m}\right)^{p} \frac{S_{ij}}{\sqrt{J_2}} \tag{4.2}$$

$$\dot{N}_m = K * k_0 * N_m * \exp(-Q/kT)\left(\sqrt{J_2} - A\sqrt{N_m}\right)^{p+\lambda} \tag{4.3}$$

$$J_2 = \frac{S_{ij} S_{ij}}{2} \tag{4.4}$$

Where $\dot{\varepsilon}_{ij}^{c}$ is creep strain rate, k is the Boltzmann constant and is equal to 8.617×10^{-5} eV K, K, k_0 and λ are material constants and given to be 3.1×10^{-4} m/N, 8.6×10^{-4} $m^{2+p}N^{-m}S^{-1}$ and 1, N_m is mobile dislocation density, A is the strain hardening coefficient which is 4 N m^{-1}, Q is the activation energy and is 2.2 eV, p is the stress exponent and is 1.1, b is the Burgers vector of the mold and is 3.8×10^{-10} m, Sij is the deviatoric stress, τeff is the equivalent shear stress, and $\sqrt{J_2}$ indicates the equivalent shear stress. The value of $\frac{dN_m}{dt}$ and $\dot{\varepsilon}_{ij}^{c}$ is set to zero when $\sqrt{J_2} - A\sqrt{N_m} \le 0$. Equations (4.1–4.4) are employed to develop a 3D transient finite element model to calculate dislocation densities generated in the ingot. The initial dislocation density for the cooling ingot is assumed to be 1×10^6 m^{-2}.

4.3 Results and Discussion

Figure 4.2b shows the variation of von Mises stress with time in normal and optimized cooling processes with HAS model. Compared with the results by CRSS model shown in Fig. 4.2a, there are two peaks and a shake in both models. The shake may be due to the effects of boundary conditions and the nonlinear physical silicon parameters. The final stress in HAS model is near 3.5 MPa in normal process while 2.2 MPa in optimized process, which means the optimized process can reduce the von Mises stress of mc-Si ingots in both models. Relative to the CRSS model, the stress goes through a long period of time to approach the stress-free state under HAS model. The reason is due to the plastic zone in such temperature field and the excessive thermal stress drive the multiplication of dislocation and therefore the stresses are partial released by plastic deformations while the stress in CRSS model increases and then decreases because it ignores the time-dependent dislocation factor. In around 150 min, the stresses in two models are all in a lower value. However, the stress in HAS model gradually

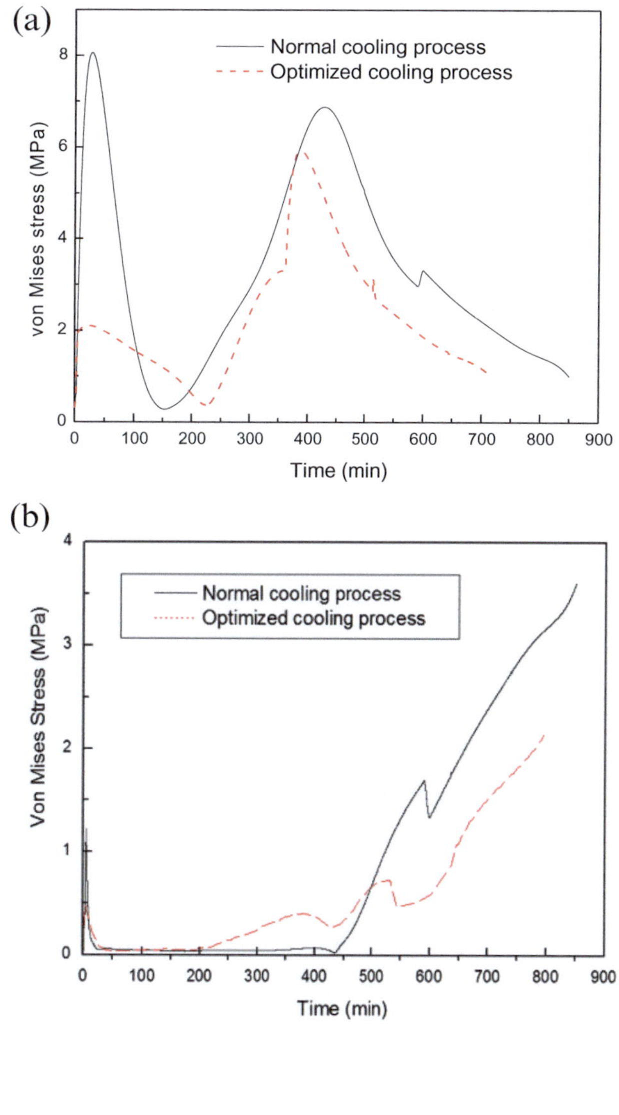

Fig. 4.2 von Mises stress of silicon ingot as a function of time in normal and optimized cooling processes with (**a**) CRSS model and (**b**) HAS model

increases until the final von Mises stress reaches the maximum, while the stress in CRSS model increases and then decreases in the final 450 min. This can be explained as follows: the CRSS model assumed the density of primary dislocations as a constant, which has ignored the equilibrium saturation of dislocation sources and the new dislocation sources from dislocation multiplication. While in the Hassen-Sumino model, the time-dependent dislocation in the brittle zone is taken into account and the excessive thermal stress mainly is the residual stress in the ingot which will increase till the stress intensity is enough to cause breakage of the ingot [7]. This indicates that the time-dependent viscoplastic deform and the creep strain effect on the final von Mises stress are significant and cannot be ignored. The CRSS model can be used to predict the final comparison of stress in different process, but HAS model are more accurate to be used in the cooling process because it can give the actual state and quantitatively predict stress during the cooling processes.

Figure 4.3b shows the variation of cumulative curves of dislocation density with time in normal and optimized cooling processes with HAS model. Like the results in CRSS model shown in Fig. 4.3a, there are two relatively steep growth processes. The two sharp rises of dislocation density correspond to the two peaks of the changing of von Mises stress shown in Fig. 4.2. In CRSS model, ρ_{d0} is the initial dislocation density at the beginning of cooling process and k is a multiplication coefficient. It can qualitatively predict the final dislocation density but it can't give the accurate value. The final dislocation density in HAS model is 45×10^6 m^{-2} in normal process while 22×10^6 m^{-2} in optimized process, which means the optimized process can reduce the dislocation density of mc-Si ingots in both models. Compared with CRSS model, the first proliferation interval time is about 10 min and then the dislocation density remains steady from about 50 to 500 min. This can be explained by the decline of the von Mises stress in Fig. 4.2b and the overall uniform thermal field at this time. Then the dislocation density increases for about 50 min and then comes to a steady state. This may explain the increase of von Mises stress in Fig. 4.2b due to no more proliferation dislocations. However, the difference of final dislocation density between two processes in HAS model is quite less than that in CRSS model. This is due to dislocation factor in the CRSS model which don't consider the creep strain factor during the cooling process. In the HAS model, the dislocation density will change with time and become lower when the stress releases. This also indicates that the HAS model are more accurate to predict the dislocation multiplication in the cooling process than the CRSS model.

Figure 4.4a, b shows the final dislocation density distribution in the ingot for normal cooling process in CRSS and HAS model. The distribution of four contour plot of dislocation density is almost the same, where the maximum dislocation

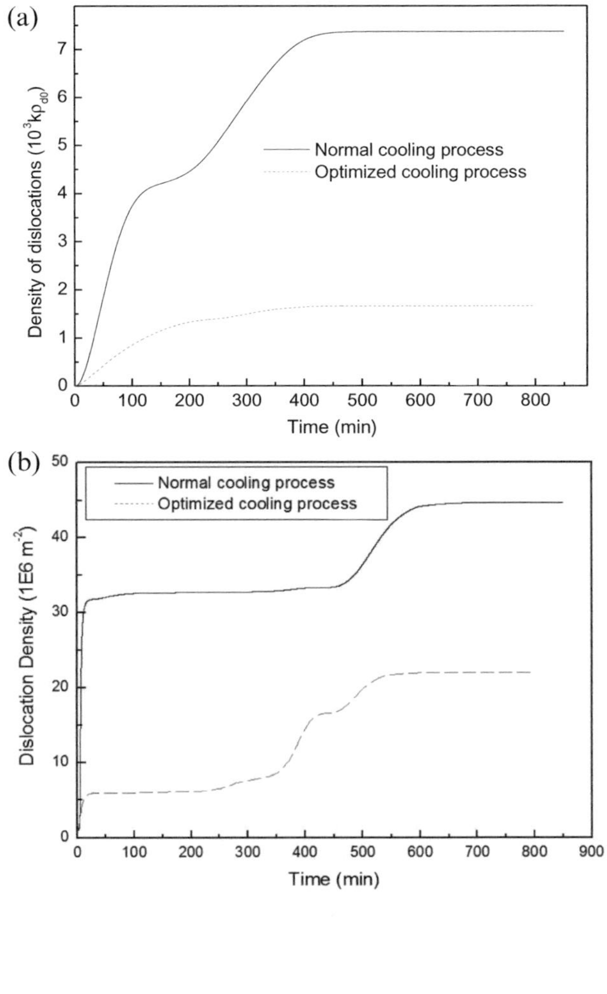

Fig. 4.3 Accumulative curves of average dislocation density in silicon ingots during cooling process with (**a**) CRSS model and (**b**) HAS model

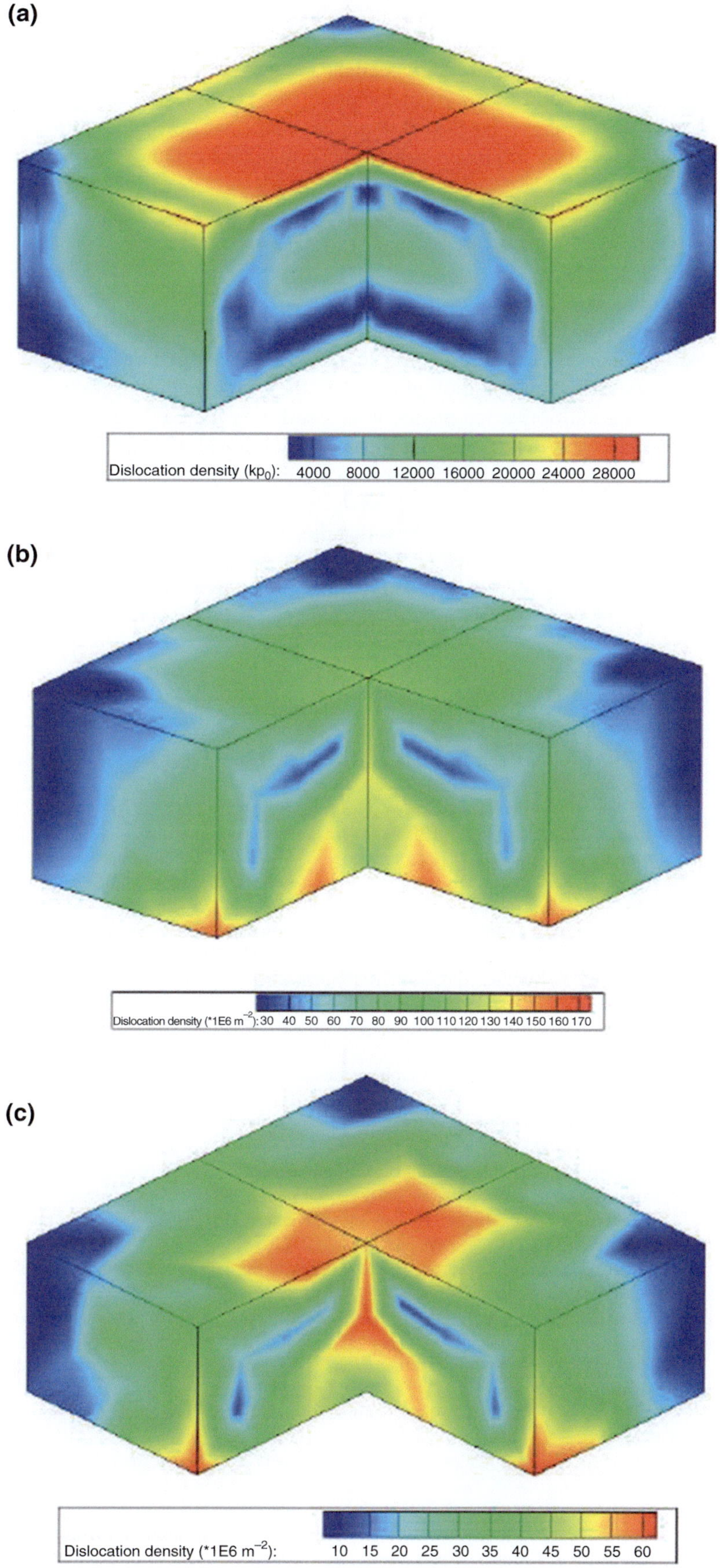

Fig. 4.4 Final dislocation density distributions of ingot during (**a**) normal cooling process with CRSS model and (**b**) normal cooling (**b**) process with HAS model (**c**) optimized cooling process with HAS model

density is on the top and the bottom of silicon ingot, while the minimum dislocation density is located in the four corners of the ingot. However, the range of final dislocation density in HAS model is quite less than that in CRSS model because of the stress release during the time-dependent viscoplastic deform process in HAS model. Figure 4.4c shows the final dislocation density distribution in the ingot for optimized cooling process in HAS model. The final maximum dislocation density in normal process is greater than the optimized process: normal (1.7×10^8 m^{-2}) > optimized (6.0×10^7 m^{-2}). Therefore, this HAS model can also be employed during the cooling process and can be used to optimize the dislocation density.

4.4 Conclusions

A three dimensional transient model to quantitatively predict dislocation multiplication in semiconductor crystals during the cooling stage of crystal growth is successfully developed based on the Haasen-Sumino viscoplastic model to coupling the microscopic dislocation multiplication to the macroscopic viscoplastic deformation. This model is employed to predict the dislocation multiplication in silicon ingots during the cooling stages of directional solidification process. The trend of von Mises stress and dislocation density distribution is similar for both HAS and CRSS models. However, the HAS model are more accurate than the CRSS model to predict the dislocation multiplication in the cooling process since CRSS is a qualitative model, where the coupling between dislocation multiplication and viscoplastic deformation is not considered. The results from HAS model also show that the optimized cooling process generate less thermal stresses and dislocation densities in the ingot than the normal cooling process, which indicates this model can also be employed by crystal growth engineers or scientists to predict the dislocation multiplication during the cooling process.

References

1. A. Poullikkas, G. Kourtis, I. Hadjipaschalis, Parametric analysis for the installation of solar dish technologies in Mediterranean regions. Renew. Sust. Energ. Rev. **14**, 2772–2783 (2010)
2. G.E. Dieter, *Mechanical Metallurgy* (McGraw-Hill, New York, 1976)
3. P. Haasen, Zur plastischen verformung von Germanium und InSb. Zeitschrift für Physik **167**, 461–467 (1962)
4. H. Alexander, P. Haasen, Dislocations and plastic flow in the diamond structure. Solid State Phys. **22**, 27–158 (1969)
5. I. Yonenaga, K. Sumino, Dislocation dynamics in the plastic deformation of silicon crystals I. Experiments. Phys. Status Solidi (a) **50**, 685–693 (1978)
6. M. Suezawa, K. Sumino, I. Yonenaga, Dislocation dynamics in the plastic deformation of silicon crystals. II. Theoretical analysis of experimental results. Phys. Status Solidi (a) **51**, 217–226 (1979)
7. I. Yonenaga, K. Sumino, K. Hoshi, Mechanical strength of silicon crystals as a function of the oxygen concentration. J Appl. Phys. **56**, 2346–2350 (1984)
8. K. Sumino, M. Imai, Interaction of dislocations with impurities in silicon crystals studied by in situ X-ray topography. Philos. Mag. A **47**, 753–766 (1983)
9. C. Tsai, O. Dillon, R. De Angelis, The constitutive equation for silicon and its use in crystal growth modeling. J. Eng. Mater. Technol. **112**, 183–187 (1990)
10. J. Moosbrugger, Continuum slip viscoplasticity with the Haasen constitutive model: application to CdTe single crystal inelasticity. Int. J. Plast. **11**, 799–826 (1995)
11. S. Pendurti, V. Prasad, H. Zhang, Modelling dislocation generation in high pressure Czochralski growth of InP single crystals: part I. Construction of a visco-plastic deformation model. Model. Simul. Mater. Sci. Eng. **13**, 249 (2005)
12. S. Nakano, X. Chen, B. Gao, K. Kakimoto, Numerical analysis of cooling rate dependence on dislocation density in multicrystalline silicon for solar cells. J. Cryst. Growth **318**, 280–282 (2011)
13. X. Chen, S. Nakano, L. Liu, K. Kakimoto, Study on thermal stress in a silicon ingot during a unidirectional solidification process. J. Cryst. Growth **310**, 4330–4335 (2008)
14. X. Chen, S. Nakano, K. Kakimoto, Three-dimensional global analysis of thermal stress and dislocations in a silicon ingot during a unidirectional solidification process with a square crucible. J. Cryst. Growth **312**, 3261–3266 (2010)
15. C. Parfeniuk, F. Weinberg, I. Samarasekera, C. Schvezov, L. Li, Measured critical resolved shear stress and calculated temperature and stress fields during growth of CdZnTe. J. Cryst. Growth **119**, 261–270 (1992)
16. G. Meduoye, D. Bacon, K. Evans, Computer modelling of temperature and stress distributions in LEC-grown GaAs crystals. J. Cryst. Growth **108**, 627–636 (1991)
17. M. Duseaux, Temperature profile and thermal stress calculations in GaAs crystals growing from the melt. J. Cryst. Growth **61**, 576–590 (1983)
18. S. Motakef, A.F. Witt, Thermoelastic analysis of GaAs in LEC growth configuration: I. Effect of liquid encapsulation on thermal stresses. J. Cryst. Growth **80**, 37–50 (1987)
19. W. Rosch, F. Carlson, Computed stress fields in GaAs during vertical Bridgman growth. J. Cryst. Growth **109**, 75–81 (1991)
20. A.S. Jordan, R. Caruso, A. Von Neida, A thermoelastic analysis of dislocation generation in pulled GaAs crystals. J. Bell Syst. Technol. **59**, 593–637 (1980)

21. A. Jordan, A. Von Neida, R. Caruso, The theoretical and experimental fundamentals of decreasing dislocations in melt grown GaAs and InP. J. Cryst. Growth **79**, 243–262 (1986)
22. O. Dillon Jr., C. Tsai, R. De Angelis, Dislocation dynamics during the growth of silicon ribbon. J. Appl. Phys. **60**, 1784–1792 (1986)
23. C. Tsai, M. Yao, A. Chait, Prediction of dislocation generation during Bridgman growth of GaAs crystals. J. Cryst. Growth **125**, 69–80 (1992)
24. C. Tsai, A. Gulluoglu, C. Hartley, A crystallographic methodology for modeling dislocation dynamics in GaAs crystals grown from melt. J. Appl. Phys. **73**, 1650–1656 (1993)
25. K. Sumino, I. Yonenaga, Dislocation dynamics and mechanical behaviour of elemental and compound semiconductors. Phys. Status Solidi (a) **138**, 573–581 (1993)
26. N. Subramanyam, C. Tsai, Dislocation reduction in GaAs crystal grown from the Czochralski process. J. Mater. Process Technol. **55**, 278–287 (1995)
27. X. Chen, S. Nakano, K. Kakimoto, 3D numerical analysis of the influence of material property of a crucible on stress and dislocation in multicrystalline silicon for solar cells. J. Cryst. Growth **318**, 259–264 (2011)
28. N. Zhou, M. Lin, M. Wan, L. Zhou, Lowering dislocation density of directionally grown multicrystalline silicon ingots for solar cells by simplifying their post-solidification processes—a simulation approach. J. Therm. Stresses **38**, 146–155 (2015)
29. N. Zhou, M. Lin, L. Zhou, Q. Hu, H. Fang, S. Wang, A modified cooling process in directional solidification of multicrystalline silicon. J. Cryst. Growth **381**, 22–26 (2013)

Chapter 5
Interaction of Shock Wave with Granular Materials

Huiyang Luo, Tingge Xu, Xuemin Wang, and Hongbing Lu

Abstract The mechanical behavior of granular materials such as beads and sand under blast wave involves complex solid–fluid coupling. In this study, a miniature shock tube is developed to investigate the interaction of shock wave with granular materials. A large-opening metal grid is used to support beads. Pressure is measured by pressure sensors as shock wave propagates. Steel beads are used for the investigation of their interaction with shock wave. A Schlieren photography system is used to observe the flow field. A high-speed camera is used to acquire video of the wave. The interaction of shock wave with beads is observed and discussed.

Keywords Shock wave • Beads • Granular materials • Schlieren photography • Interaction

5.1 Introduction

The high-speed deformation behavior of granular materials under blast wave is not well understood. For modeling the complex mechanisms of deformation, numerical simulations can shed light on the understanding of the fundamental mechanisms of deformation of granular materials. These simulations can capture the effect of solid/fluid interactions on the mechanical behavior [1, 2]. The actual grain geometry is irregular, the grains are often represented by spheres or ellipsoids as model granular materials in simulations [3–5]. Current computational modeling to simulate blast and ejecta in soils relies heavily on the continuum approaches, to couple mesoscale simulations with the mechanical response of discrete sand grains in areas of interest [6]. The constitutive data for sand/soil under high strain rates have been investigated by a modified split Hopkinson pressure bar (SHPB). Sand was investigated under triaxial loading through active hydraulic-pressure confinement on a modified SHPB at high strain rates [7]. Effects of particle size and moisture were determined on the compressive behavior of sand at high strain rates [8–12]. The stress–strain relationship of sand at high strain-rates was input for soil mechanics modeling under blast [13]. A review was given in this area [14]. However, the granular materials under blast wave were not well simulated and modelled due to the lack of experimental data for the validation.

To simulate blast wave and their effects, a shock tube is often used to replicate the direct blast waves. Results from shock tube experiments can be further analyzed to validate numerical model of the response of a material subjected to a blast wave. Shock wave interaction with solid sphere can provide understanding of shock dynamics. The interaction generates non-linear aerodynamic behavior involving unsteady drag force on the ball, complex wave reflections such as the Mach reflection, and attenuation of shock waves resulting from the wave interactions with vortices in the wake of each sphere. The drag coefficient of a sphere placed in a stationary/non-stationary flow has been studied experimentally over a wide range of Reynolds numbers in subsonic and supersonic flows [15–19]. Optical methods such as Schlieren photography, holographic interferometry, and other shadowgraph observations were often applied to visualize the various shock waves passing through solid spheres, and high-speed images were often captured by high-speed camera [20, 21].

The unsteady drag force on sphere in a shock tube has received attention recently. An experiment on unsteady two-phase flow was conducted in a vertical shock tube, and shadowgraph images were used to record particle trajectories [21]. A large sphere was placed in the path of a vertical shock tube, its acceleration was measured by an embedded accelerometer [22]. The dynamic drag coefficient of a sphere was investigated numerically and experimentally [16]. Experiments for the aerodynamic processes involved in the interaction between shock waves and stationary-fixed solid spheres, to determine the

---

H. Luo • T. Xu • X. Wang • H. Lu (✉)
Department of Mechanical Engineering, The University of Texas at Dallas, Richardson, TX 75080, USA
e-mail: hongbing.lu@utdallas.edu

B. Antoun (ed.), *Challenges in Mechanics of Time Dependent Materials, Volume 2*,
Conference Proceedings of the Society for Experimental Mechanics Series, DOI 10.1007/978-3-319-22443-5_5

effect of the sphere array, size, interval distance, incident Mach number, on the shock wave attenuation [15]. Shock-tube was also used to investigate the dynamics of gas-particle mixtures under shock wave; the results indicated that particles provide resistance to gas flow resulting from interphase drag, thus reducing shock strength and velocity [23]. In another investigation, balls were suspended inside shock tube; they were visualized using a high-speed Schlieren shadowgraph system, to measure drag coefficient [17], and to investigate shock propagation in a shock-wave/turbulent boundary-layer interaction with beads [20]. Particle drag coefficients were measured and the elevated particle drag coefficients increased with compressibility [18, 19].

In this study, a miniature vertical shock tube has been developed to generate shock wave to interact with steel beads supported by a metal grid. High-speed Schlieren imaging was used for visualization and the pressure was measured by pressure sensors, to investigate the interaction of shock wave with beads.

5.2 Experimental Aspects

Shock tube. The shock tube used in this study consists of a high pressure driver section and a low pressure driven section separated by a membrane diaphragm, and a testing section in the downstream at the top. A sudden burst of diaphragm under high pressure driver section generates a blast wave, which propagates to reach the test section to interact with beads. A miniature vertical shock was developed, its schematic diagram is shown in Fig. 5.1. The tube was made of steel, with 1.07″ outer diameter and 0.75″ inner diameter. The driver section had a length of 9 ft bent into U-shape. The tubes on the left and right were each 2 ft long, and the tube at the bottom tube was 5 ft long. This geometry was selected in order to give convenience in the installation of the optical system. A metal grid supported by a ring was attached to the exit of the 6″ long driven section; it was used to support 12 steel balls, which each had 3/16″ diameter. Four pressure sensors (PCB 101A06), connected to a four-channel signal conditioner (PCB 482C), were used to measure pressure as the shock wave traveled through. The pressure sensor head had 3/8″ diameter. HBM Genesis 5i oscilloscope (15-bit resolution, 25 ms/s sampling rate, 5 MHz bandwidth) was used to acquire the pressure data. A 0.004″ thick polyester film (McMaster # 8567K42) was used as diaphragm.

Schlieren system. The rectangle region with cross section of 1.5″ by 1.0″ above the beads at the top center of Fig. 5.1 was chosen as the viewing area, which was acquired by a high-speed camera (Photron SA6). An Edmund Schlieren system (Edmund Optics, #71-014) used 4.25″ diameter mirrors with 45″ focal length. The system contained two protected

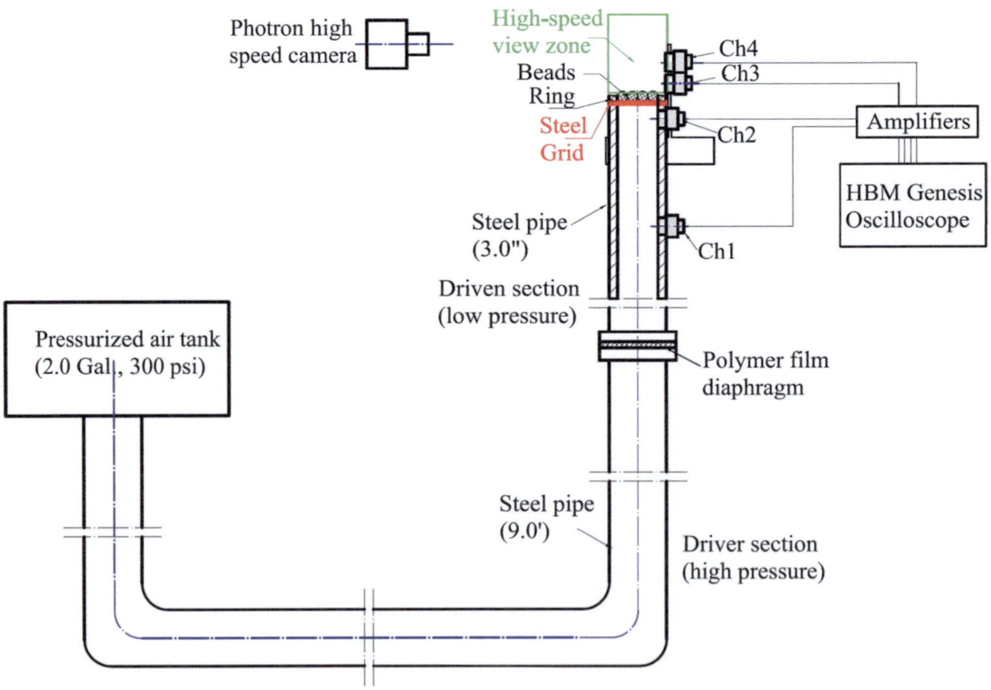

Fig. 5.1 Modified miniature vertical shock tube setup

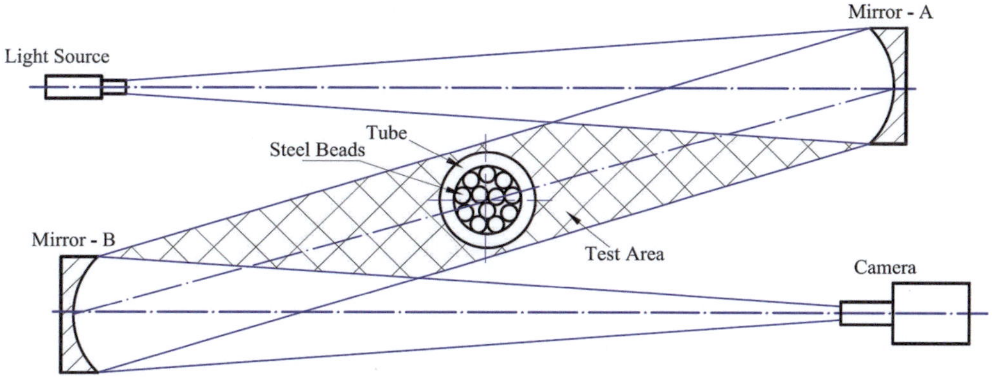

Fig. 5.2 Schematic Schlieren setup on modified shock tube

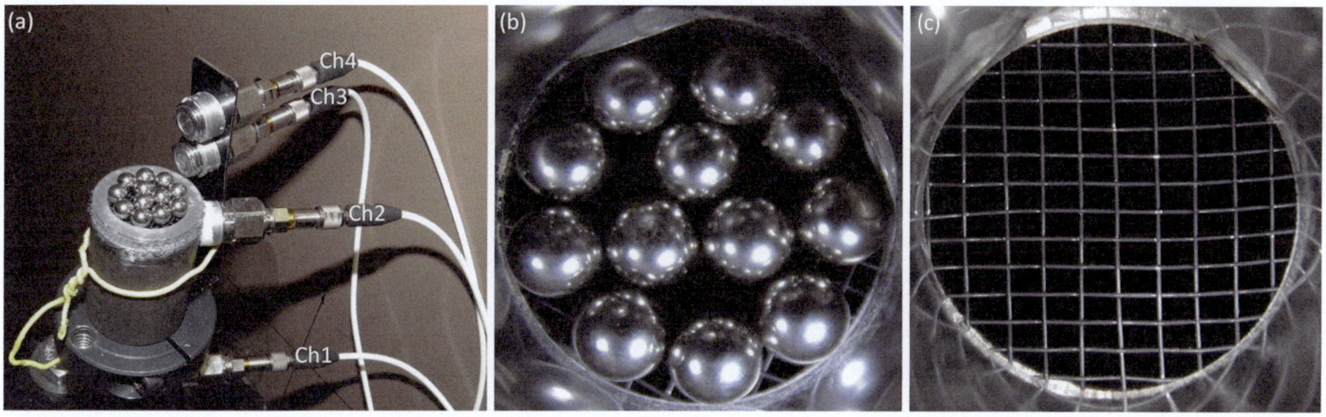

Fig. 5.3 Testing section of shock tube. (**a**) A picture showing the test section; four PCB pressure sensors are shown. (**b**) 12 pieces of steel beads on the metal grid. (**c**) Metal grid

aluminized front-surface spherical f/10 mirrors with 1/8 wave flatness; they were overcoated with silicon monoxide and mounted on isolated metal stands.

As shown in Fig. 5.2, mirror (A) acted as a collimator. A high-intensity light source at its focus point was reflected to form a parallel beam. At an appropriate location the second mirror (B) reflected the parallel beam to focus it onto a high-speed camera. The test section contained 12 steel beads; it was placed in the center of the Schlieren system. The two mirrors were installed on stands, which were not affected by the shock wave to images.

Test section. The test section was open for ease of observation under the Schlieren system. 12 pieces of smooth steel beads were placed on the metal grid mounted at the end of the driven tube; the beads were surrounded by a 1/8″ thick polycarbonate ring (1.0″ OD and 0.75″ ID), as shown in Fig. 5.3a. Four PCB pressure sensors, they were marked with Ch 1–4 in Fig. 1, were placed vertically. Ch 1 and 2 sensors were mounted on the tube of driven section. Sensors of Ch 3 and 4 were supported by a steel plate mounted on the shock tube, in order to measure pressure at the edge in the path of shock wave exited from the 0.75″ diameter tube. Sensor Ch 2 is just below the metal grid and Ch 1 is 2.5″ far above from Ch 1. Sensor Ch 3 was slightly above the beads and Ch 4 was next to Ch 3, separated by 3/8″ distance. The 0.75″ ID tube was just big enough to fit the 12 beads (Fig. 5.3b). The metal grid had 0.008″ steel wire and 66 % opening area (Fig. 5.3c).

Shock Experiments. Since a metal grid was used to support the beads, the effect of grid on shock wave was also examined separately. Several experiments were thus conducted. At first, there were no beads or grid used, as shown schematically in Fig. 5.4a, to determine the effect of U-shape driver section on pressure profile. Next, the supporting metal grid and surrounding ring were mounted (Fig. 5.4b), to determine the effect of grid. Finally, beads were placed on the grid (Fig. 5.4c), to investigate the interaction of shock wave with beads. The pressurized driver tube was filled with compressed air at 300 psi pressure. In each test, when the film diaphragm ruptured, the shock wave propagated to reach the testing section, a high speed camera was used to acquire Schlieren shadowgraphs; the oscilloscope was triggered to acquire the signals from the four pressure sensors.

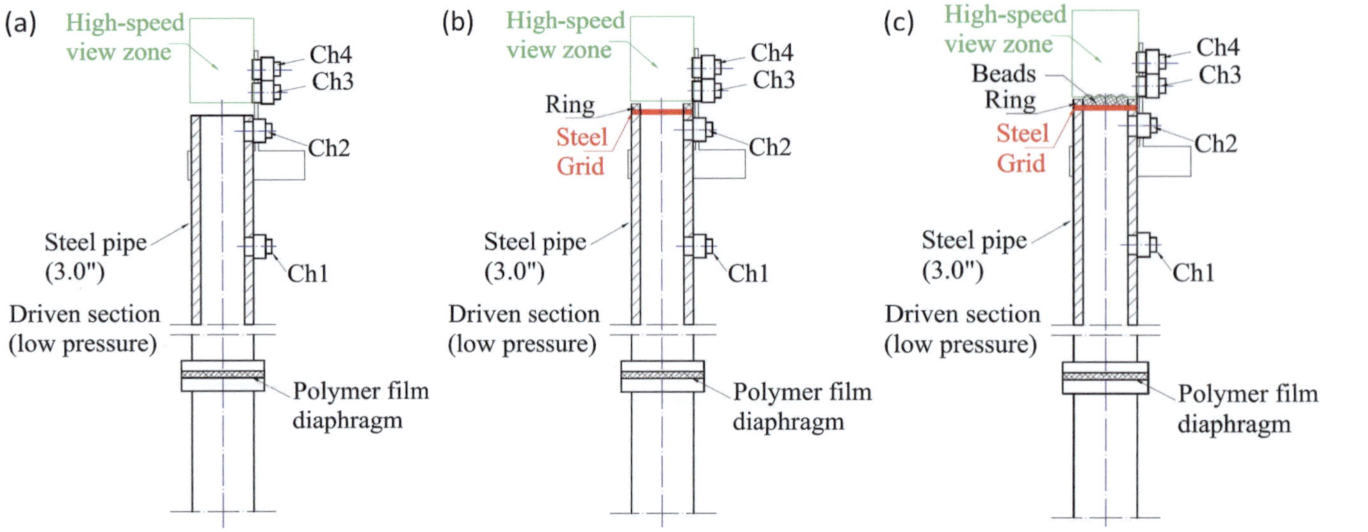

Fig. 5.4 Schematic setup of test section of shock tube. (**a**) Without sample and grid. (**b**) With grid and without sample. (**c**) With grid and sample

5.3 Results and Discussion

Once the diaphragm ruptures, a shock wave front propagates into the driven section at a constant shock speed. When the shock wave reaches the test section, a rarefaction wave travels back into the driven and driver sections. Then it reflects back to the closed end of the driver section, forming the second shock wave, resulting in the second loading on the testing section. Thus, the length of driver section is critical to provide enough shock duration before the second shock wave arrives. Figures 5.5, 5.6 and 5.7 show the pressure histories measured by four pressure sensors at 300 psi driver pressure without sample and (Fig. 5.4a), with grid and without sample (Fig. 5.4b), and with sample and grid (Fig. 5.4c), respectively.

Figure 5.5 shows that when there were no grid and sample, all shock wave passed through, with negligible reflection. The shock wave speed was measured as 620 m/s between Ch 1 and 2. Under 300 psi pressure in the driver section, the shock wave peak pressure of 53 psi was measured by Ch 1 sensor, and 54 psi by Ch 2 sensor, respectively. After the shock wave propagated into the open atmosphere, it attenuated to 49 psi and 39 psi, at distances of 0.32″ and 0.77″ above the exit of the driven section, respectively. This result indicates that the U-shape driver section has negligible effect on shock wave.

Figure 5.6 shows that when the grid was used without sample, shock wave mostly passed, with some reflection of the rarefaction wave. The shock wave speed was kept the same while the rarefaction wave speed was measured as 120 m/s between the second jumps measured by Ch 1 and 2 pressure sensors. The shock peak was 49 psi as measured by Ch 1 sensor, and 50 psi by Ch 2 sensor, respectively, close to those in Fig. 5.5. As the rarefaction waves arrived, the Ch 1 had the second peak at 34 psi, while Ch 2 had higher value at 71 psi, respectively. After the shock wave propagated into the open atmosphere, it attenuated into 29 psi as measured by Ch 3, and 24 psi by Ch 4, respectively. The wave shape looked like the Friedlander wave; it dropped rapidly to negative value from the step peak, then rising slowly. The results indicated that the 66 %-open sieve did not have significant effect on shock wave.

Figure 5.7 shows the situation when both grid and sample were used, the rarefaction wave was observed to be significant. The shock wave speed was kept the same while the rarefaction wave speed increased to 230 m/s. The shock peak has 52 psi, as measured by Ch 1 sensor, and 54 psi as measured by Ch 2 sensor, respectively, close to those in Figs. 5.5 and 5.6. After the shock wave interacted with grid and bead, it attenuated to 19 psi, as measured by Ch 3 and 20 psi by Ch 4, respectively. As the rarefaction waves arrived, Ch 1 recorded the second peak at 80 psi, while Ch 2 recorded 152 psi, nearly twice of the pressure for the case with sample and without grid (Fig. 5.6). Except for Ch 3, which was placed at the center location of beads, the wave pressure increased slowly after Friedlander wave. The second shock loading started at ~1.6 ms, indicated by the third small peak while the shock duration was determined as 1.1 ms. The rapid fluctuation in Ch 3 was likely due to the movement of some beads placed close to the sensor, not due to the actual shock fluctuation. This indicated the beads were accelerated by shock wave and started to move above and thus the drag force was induced, leading to attenuation of the shock wave. Additional evidence will be provided by the Schlieren shadowgraph below since it shows field variation of the refractive index of air due to change of pressure or temperature under shock wave.

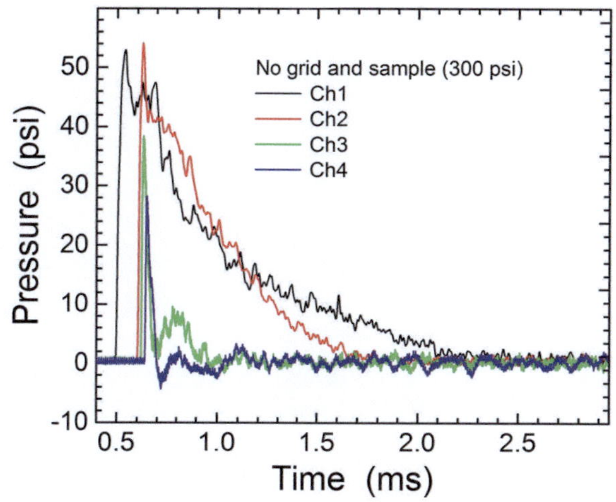

Fig. 5.5 Shock wave signals at 300 psi driver pressure without sample and grid

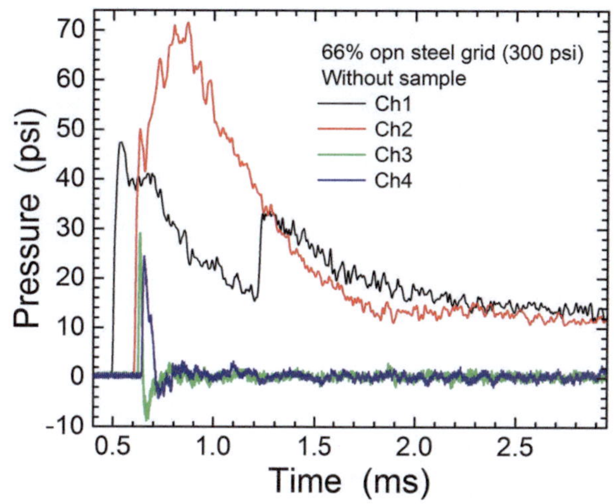

Fig. 5.6 Shock wave signals at 300 psi driver pressure for grid without sample

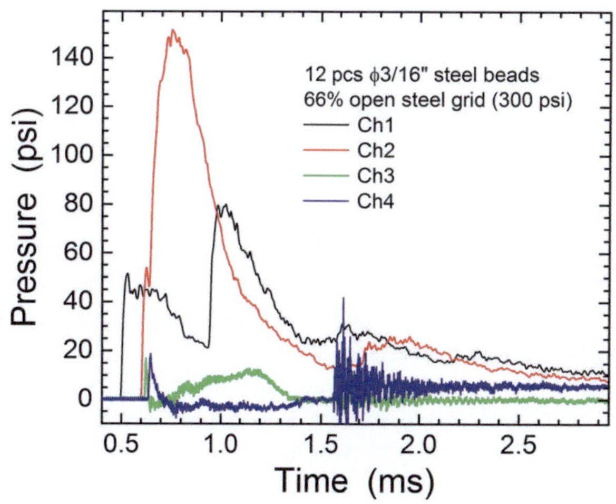

Fig. 5.7 Shock wave signals at 300 psi driver pressure with both grid and sample

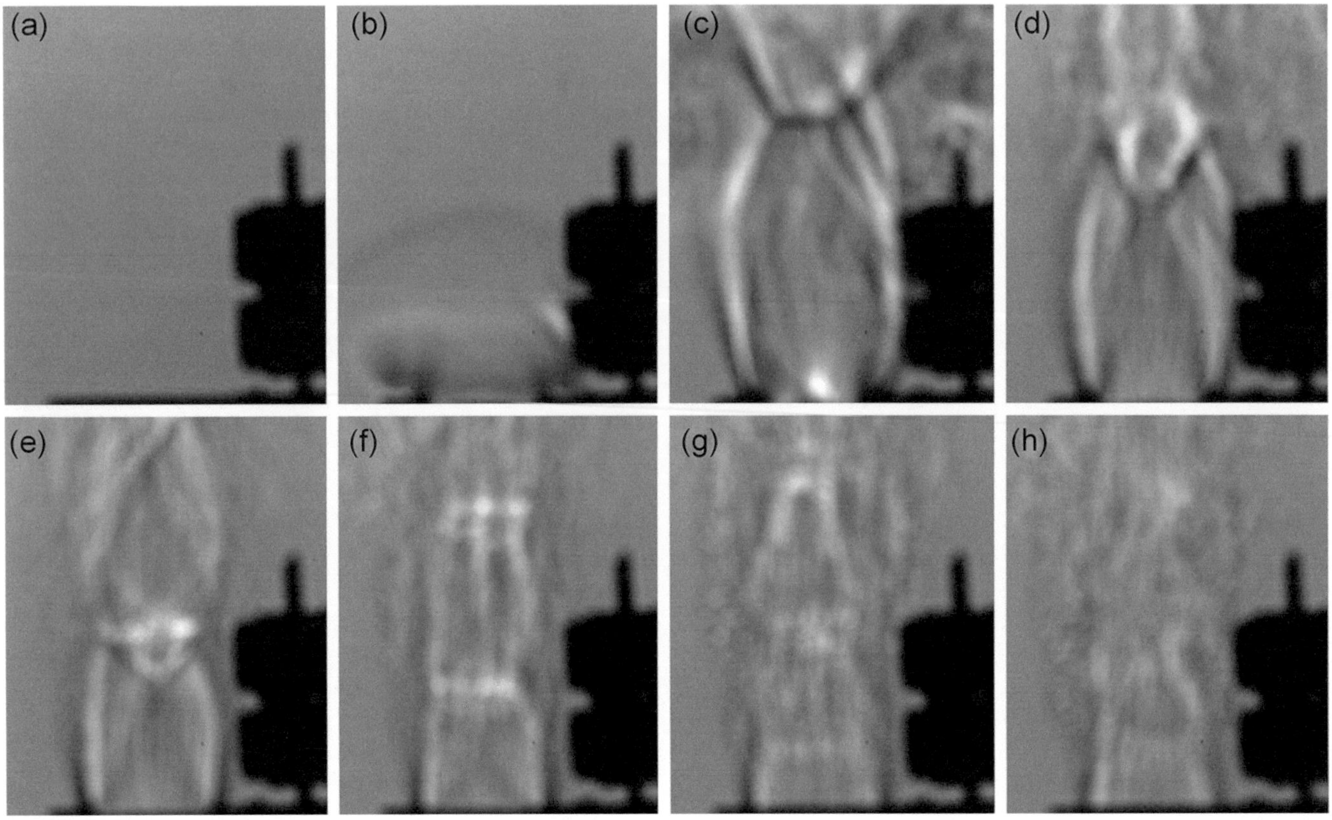

Fig. 5.8 Schlieren images without both grid and sample (5 k fps). (**a**) t=0 ms; (**b**) t=0.2 ms; (**c**) t=0.4 ms; (**d**) t=0.6 ms; (**e**) t=0.8 ms; (**f**) t=1.0 ms; (**g**) t=1.2 ms; (**h**) t=1.4 ms

Schlieren photography records a set of lighter and darker patches corresponding to positive and negative fluid density gradients. When a knife-edge is used, the system is generally referred to as a Schlieren system, which measures the first derivative of density in the direction of the knife-edge. If a knife-edge is not used, the system is generally referred to as a shadowgraph system, which measures the second derivative of density [24, 25]. When the fluid flow is uniform, the image will be steady, any turbulence will cause distorted images. The Schlieren apparatus shows changes in the index of refraction in the air since the density changes as the pressure and temperature change. Schlieren photography is a visual process used to photograph the flow of fluids of varying density. It is widely used in aeronautical engineering to photograph the flow of air around objects. The basic optical Schlieren system uses light from a single collimated source in behind, to target object. Variations in refractive index caused by density gradients in the fluid/gas distort the collimated light beam. This distortion creates a spatial variation in the intensity of the light, visualized directly with a shadowgraph system. A high-speed camera Photron SA6 was used to record the Schlieren images. The frame rate was chosen at 5000 fps, which gives frame interval as 200 μs.

Figure 5.8 shows Schlieren shadowgraphs without grid or sample. Figure 5.8a shows the initial status without shock wave, and Fig. 5.8b–h show the successive images at intervals of 200 μs. The same arrangement was used for Figs. 5.9 and 5.10. When no shock wave was present, the illumination was flat, bright and uniform (Fig. 5.8a). The image had resolution of 512×640 pixels^2, each pixel represents 3.41×10^{-3} inch actual length. Note that two black blocks on the right side were Ch 3 and 4 sensors. Since the shock duration was 1.1 ms, Figs. 5.8b–f show the first shock loading, and Fig. 5.8g, h show the second shock loading, which was not the topic of interest; they were given here just for comparison. At 200 μs when shock wave existed the nozzle, a mushroom-like shock front was seen (Fig. 5.8b); this was also observed in Figs. 5.9b and 5.10b. At 400 μs, the boundary of shock wave was clearly observed (Fig. 5.8c), similar to the shape of a candle flame. A steady upstream laminar flow was clearly seen at the central region. At 600 μs, 800 μs and 1000 μs, the boundary became smaller.

Figure 5.9 shows Schlieren shadowgraph images with grid and without sample. Unlike Figs. 5.8b–f, the stratosphere changed into cross-like flow on both the left and the right, somewhat like a 45° inclined grid-shape. This was due to the grid effect since the metal grid changed the shock wave path slightly. Figure 5.10 shows Schlieren shadowgraphs in the case with

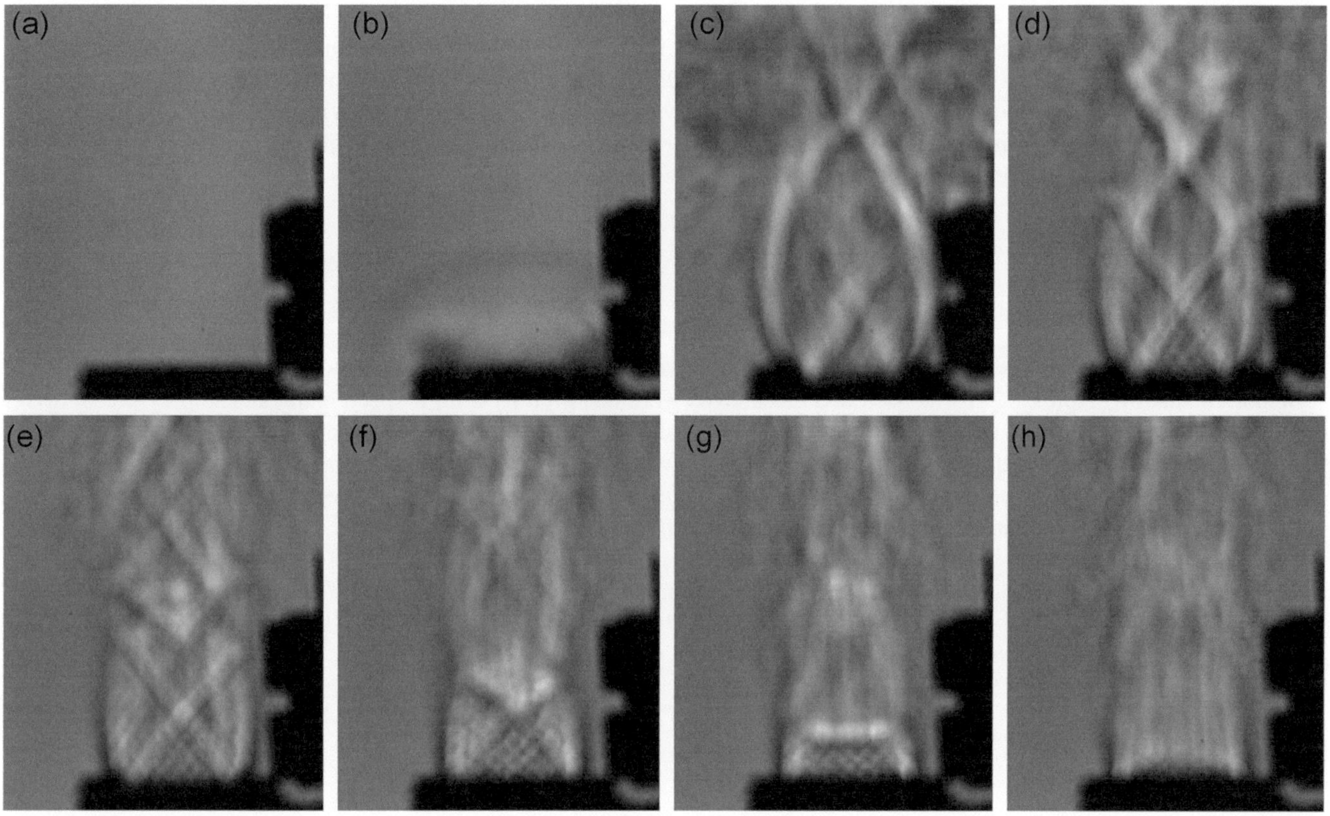

Fig. 5.9 Schlieren images with grid without sample (5 k fps). (**a**) t=0 ms; (**b**) t=0.2 ms; (**c**) t=0.4 ms; (**d**) t=0.6 ms; (**e**) t=0.8 ms; (**f**) t=1.0 ms; (**g**) t=1.2 ms; (**h**) t=1.4 ms

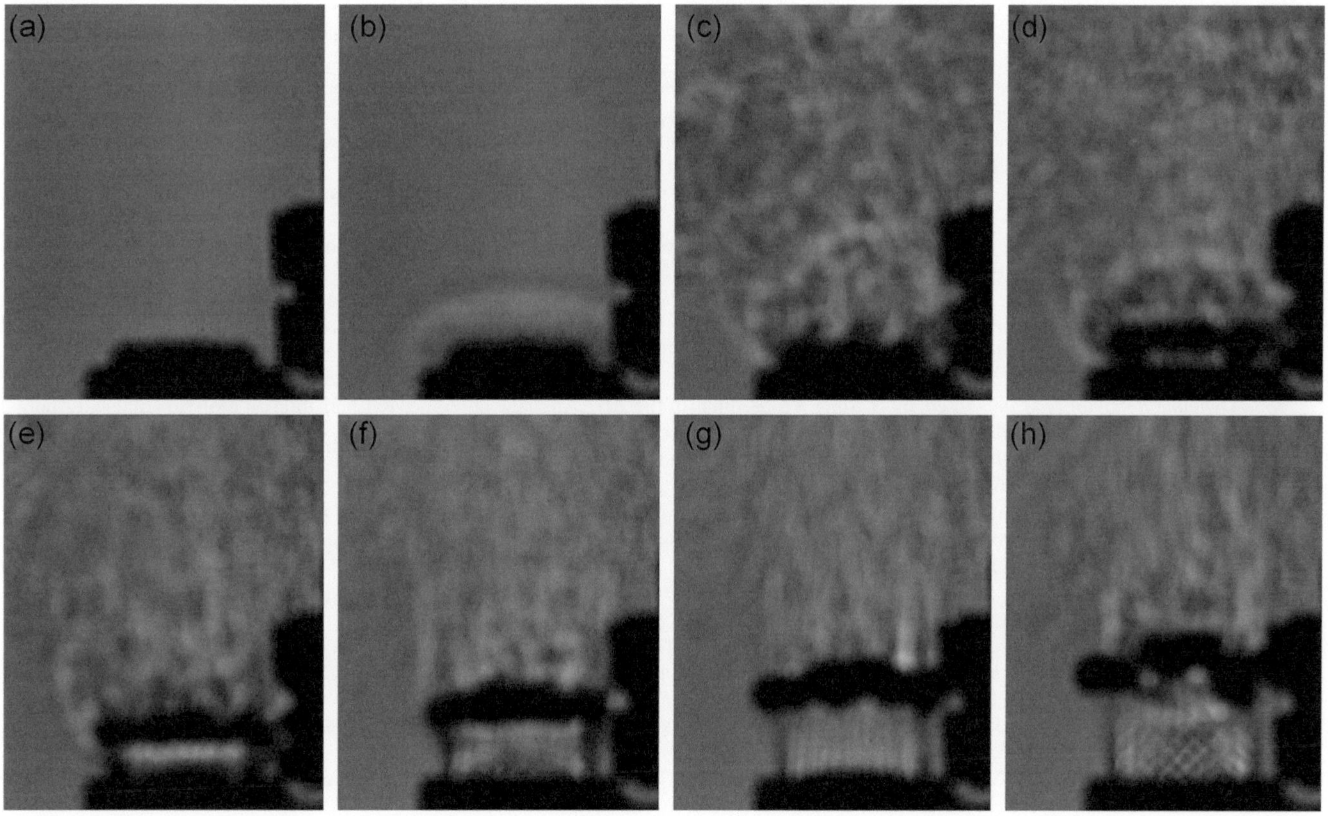

Fig. 5.10 Schlieren images with both sample and grid sample (5 k fps). (**a**) t=0 ms; (**b**) t=0.2 ms; (**c**) t=0.4 ms; (**d**) t=0.6 ms; (**e**) t=0.8 ms; (**f**) t=1.0 ms; (**g**) t=1.2 ms; (**h**) t=1.4 ms

both grid and sample. In Fig. 5.10b, the initial shock wave was kept similar to the case without grid and sample. Unlike Figs. 5.9b–f, the stratosphere changed into turbulent flow as shown in Fig. 5.10c. At 600 μs in Fig. 5.10d, the black disks on the bottom were the steel beads, they started to move upward. In Figs. 5.10c–f, this turbulent flow only happened after the shock had passed through beads, but they were laminar flow before reaching the beads. When the second shock loading arrived (Fig. 5.10h), it had formed the similar 45° cross flow behind the beads. Since the Schlieren images showed changes in the index of refraction, these images can be analyzed further in modeling.

5.4 Conclusion

A modified miniature vertical shock tube was developed to investigate the interaction of shock wave with granular materials. A Schlieren shadowgraph system was used and thc images were acquired by a high-speed camera. The pressure in driver tube was 300 psi. 12 pieces of steel beads were used for investigation of interaction of chock wave with granular materials. Images of Schlieren shadow show interaction of shock wave with steel beads. The 66 % open steel screen grid shows some drag reduction on shock wave. Future work is planned to focus on multi-layers of beads under shock wave, interaction with large beads, sands/soil, etc. The data can be analyzed further for calibration of simulations.

Acknowledgement We acknowledge the support of ONR Multidisciplinary University Research Initiative program (MURI) grant N00014-11-1-0691, and NSF under CMMI-1031829 and ECCS- 1307997. Lu also thanks the Louis A. Beecherl Jr. Chair at the University of Texas at Dallas for additional support.

References

1. D.J. Benson, V.F. Nesterenko, F. Jonsdottir, M.A. Meyers, Quasistatic and dynamic regimes of granular material deformation under impulse loading. J. Mech. Phys. Solids **45**, 1955–1999 (1997)
2. P. Fu, Y.F. Dafalias, Quantification of large and localized deformation in granular materials. Int. J. Solids Struct. **49**, 1741–1752 (2012)
3. L.L. Ragione, J.T. Jenkins, The initial response of an idealized granular material. Proc. Math. Phys. Eng. Sci. **463**, 735–758 (2007)
4. J.N. Roux, Geometric origin of mechanical properties of granular materials. Phys. Rev. E **61**, 6802–6836 (2000)
5. W.L. Cooper, B.A. Breaux, Grain fracture in rapid particulate media deformation and a particulate media research roadmap from the PMEE workshops. International Journal of Fracture **162**, 137–150 (2010)
6. R. Regueiro, R. Pak, J. McCartney, S. Sture, B. Yan, Z. Duan, J. Svoboda, W. Mun, O. Vasilyev, N. Kasimov, E. Brown-Dymkoski, C. Hansen, S. Li, B. Ren, K. Alshibli, A. Druckrey, H. Lu, H. Luo, R. Brannon., C. Bonifasi-Lista, A. Yarahmadi, E. Ghodrati, J. Colovos, ONR MURI project on soil blast modeling and simulation, *Chapter 42 in Dynamic Behavior of Materials*, ed. by B. Song et al. The Conference Proceedings of the Society for Experimental Mechanics Series, **1**, 341–353 (2014)
7. B.E. Martin, E. Kabir, W. Chen, Undrained high-pressure and high strain-rate response of dry sand under triaxial loading. Int. J. Impact Eng. **54**, 51–63 (2013)
8. H. Luo, H. Lu, W.L. Cooper, R. Komanduri, Effect of mass density on the compressive behavior of dry sand under confinement at high strain rates. Exp. Mech. **51**(9), 1499–1510 (2011)
9. H. Lu, H. Luo, W.L. Cooper, R. Komanduri, Effect of particle size on the compressive behavior of dry sand under confinement at high strain rates, *Chapter 67 in Dynamic Behavior of Materials*. Proceedings of the 2012 Annual Conference & Exposition on Experimental and Applied Mechanics, Conference Proceedings of SEM Series C,**1**, 523–530 (2013)
10. H. Luo, W.L. Cooper, H. Lu, Effect of moisture on the compressive behavior of dry sand under confinement at high strain rates, in *Chapter 46 in Dynamic Behavior of Materials* ed. by B. Song. Proceedings of the 2013 Annual Conference & Exposition on Experimental and Applied Mechanics, Conference Proceedings of SEM series, Vol. 1, (Springer, 2014), pp. 381–388
11. H. Luo, W.L. Cooper, H. Lu, Effects of particle size and moisture on the compressive behavior of dense Eglin sand under confinement at high strain rates. Int. J. Impact Eng. **65**, 40–55 (2014)
12. H. Luo, Y. Du, Z. Hu, H. Lu. High-strain rate compressive behavior of dry mason sand under confinement. *Chapter 46 in Dynamic Behavior of Materials* , ed. by B. Song. Proceedings of the 2014 Annual Conference on Experimental and Applied Mechanics, Conference Proceedings of the Society for Experimental Mechanics Series, Vol. 1, (Springer, 2015) pp. 381–388
13. W. Higgins, T. Chakraborty, D. Basu, A high strain-rate constitutive model for sand and its application in finite-element analysis of tunnels subjected to blast. Int. J. Numer. Anal. Methods Geomech **37**(15), 2590–2610 (2013)
14. M. Omidvar, M. Iskander, S. Bless, Stress–strain behavior of sand at high strain rates. Int. J. Impact Eng. **49**, 192–213 (2012)
15. H. Shi, K. Yamamura, The interaction between shock waves and solid spheres arrays in a shock tube. Acta Mech. Sinica **20**, 219–227 (2004)
16. M. Sun, T. Saito, K. Takayama, H. Tanno, Unsteady drag on a sphere by shock wave loading. Shock Waves **14**, 3–9 (2004)
17. G. Jourdan, L. Houas, O. Igra, J.L. Estivalezes, C. Devals, E.E. Meshkov, Drag coefficient of a sphere in a non-stationary flow: new results. Proc. Math. Phys. Eng. Sci. **463**, 3323–3345 (2007)
18. J.L. Wagner, S.J. Beresh, S.P. Kearney, B.O. Pruett, E. Wright, Shock tube investigation of unsteady drag in shock-particle interactions. *41st AIAA Fluid Dynamics Conference and Exhibit.* (Honolulu, HI, AIAA), 27–30 June 2011

19. J.L. Wagner, S.J. Beresh, S.P. Kearney, W.M. Trott, J.N. Castaneda, B.O. Pruett, M.R. Baer, A multiphase shock tube for shock wave interactions with dense particle fields. Exp. Fluids **52**, 1507–1517 (2012)
20. D. Estruch, N.J. Lawson, D.G. MacManus, K.P. Garry, J.L. Stollery, Measurement of shock wave unsteadiness using a high-speed Schlieren system and digital image processing. Rev. Sci. Instrum. **79**, 126108 (2008)
21. X. Rogue, G. Rodriguez, J.F. Haas, R. Saurel, Experimental and numerical investigation of the shock-induced fluidization of a particles bed. Shock Waves **8**, 29–45 (1998)
22. H. Tanno, K. Itoh, T. Saito, A. Abe, K. Takayama, Interaction of a shock with a sphere suspended in a vertical shock tube. Shock Waves **13**, 191–200 (2003)
23. K. Chojnicki, A.B. Clarke, J.C. Phillips, A shock-tube investigation of the dynamics of gas-particle mixtures: implications for explosive volcanic eruptions. Geophys. Res. Lett. **33**, L15309 (2006)
24. G.S. Settles, *Schlieren and Shadowgraph Techniques: Visualizing Phenomena in Transparent Media* (Springer, Berlin, 2001)
25. P.K. Panigrahi, K. Muralidhar, Laser schlieren and shadowgraph, in *Chapter 2 in Schlieren and Shadowgraph Methods In Heat And Mass Transfer*, Springer Briefs in Thermal Engineering and Applied Science, (Springer, 2012), pp. 23–46

Chapter 6
Characterisation of Viscoelastic Material Properties During Curing Processes

S. Saseendran, M. Wysocki, and J. Varna

Abstract The present contribution is toward systematic characterisation of the thermo-viscoelastic properties of a curing epoxy resin system. Characterising the viscoelastic solid behaviour is performed using a dynamic mechanical analyser. The aim of this work is to investigate the dependence of the viscoelastic response on time, temperature and degree of cure and to derive a model that covers the dependency of the relaxation modulus on all three factors and also to investigate how various factors would influence each other in the overall evolution of the relaxation modulus. In particular, we investigate the linearity between the three factors above. To summarize, the results indicate that these three parameters indeed obey a linear relationship.

Keywords Residual stress • Shape distortion • Viscoelasticity • Curing • DMTA

6.1 Introduction

Composite materials are attractive for structural applications due to their lightness and in particular their high strength to weight ratios. This is of particular importance in the transportation sector, for instance where low weight can lead to fuel savings or to lightweight ships and aircrafts for increased payload. However to enable the aforementioned trend it is necessary to increase understanding of the physics in order to facilitate the development composite products. Traditionally, composites have been developed using trial and error methods resulting in an expensive development process. This is in particular true for the development of manufacturing processes, where many experimental loops are necessary before an efficient manufacturing process is established [1]. Thus new modelling knowledge and techniques are necessary to reduce developmental costs incurred by developing new components. One such area considered here is the prediction of manufacturing induced imperfections such as residual stresses and shape distortion based on viscoelastic behavior of resins during cure.

Experimental studies have been performed in the past in order to characterise development of viscoelastic properties during cure. Eom et al. [2], characterised in a rheometer the entire curing process with various curing temperatures. The obtained values relate to the shear storage modulus development over the curing process. Eom et al. have been able to determine that the shift factors have a linear relationship with the cure temperatures for normal time-temperature superposition. Yeong et al. [3] have ascertained that since most of the residual stresses develop at post-gelation it is ideal to look for ways to model stress relaxation at such post-gelation conditions. Their effort focussed primarily on the stress relaxation behaviour of epoxy resin samples that have been cured to a particular degree of cure using several different temperatures to cure. O'Brien et al. [4] have performed experiments with neat epoxy resins with a single cure temperature and have found that the cure state has no effect on the elastic response of the material. O'Brien et al. have also noted that the peak equilibrium relaxation modulus varies significantly between the fully cured and uncured states and hence hinting at a possible interdependence of the equilibrium modulus and degree of cure. Further, in the work by Thorpe and Pousartip [5], an attempt to develop a model based on a carbon fibre cured epoxy system was carried out. Characterisation was performed in the solid state in a DMA and in the liquid state for a neat resin in a rheometer. The findings have concluded that a thermo-rheologically simple time-temperature superposition is insufficient to characterise viscoelastic behaviour. The study has

S. Saseendran (✉) • M. Wysocki
Swerea SICOMP AB, Piteå 94332, Sweden
e-mail: sibin.saseendran@swerea.se

J. Varna
Luleå Technical University, Luleå 97187, Sweden

B. Antoun (ed.), *Challenges in Mechanics of Time Dependent Materials, Volume 2*,
Conference Proceedings of the Society for Experimental Mechanics Series, DOI 10.1007/978-3-319-22443-5_6

also concluded that the glassy state modulus is independent of the degree of cure, a behaviour which is also confirmed by Sadeghinia et al. [6] alternatively in the shear modulus domain. This is in contradiction to O'Brien et al. and the experiments performed by the same. Suzuki et al. [7] have reported that the form of a master curve generated through time-temperature superposition principles is dependent on the cure temperature and time and both factors are to be considered separately. However in their work, Suzuki et al. have not made an attempt to study cure kinetics and subsequent comparisons with the cure state of the resin. In the work by Simon et al. [8], a model for the calculation of the evolution of shear storage modulus that uses inputs of cure kinetics was formulated. This model incorporates time-temperature superposition and time-cure state superposition in thermosets. An attempt to model a cure dependent model for the evolution of the storage modulus in a high temperature epoxy system based on a modified form of the Kohlrausch-Williams-Watts (KWW) equations has been performed in the work by Zarrelli et al. [9]. Adolf and Martin [10] have done preliminary assessment of the influence of cure on viscoelastic properties by developing a time-cure superposition method. This method uses horizontal and vertical shifting of DMA data to arrive at master curves and the approximation uses power law fitting rather than conventional Prony series fitting. The vertical shift accounts for the change in the equilibrium modulus with changing degree of cure which has never been accounted for in some other similar works. They have worked on the shear modulus domain and have not conducted an in-depth analysis of the cure profile of the material under test. They have ascertained that viscoelastic functions at differing extents of reaction should superpose if the appropriate vertical and horizontal shifts are applied to the DMA data.

In the present development we assume the material to depend, beside frequency, on the following parameters only: temperature (T), curing temperature (T_c) and the curing time (t_c). Therefore, we intend to investigate the stress–strain relationship of the form

$$\sigma = h_e\left(T_c, t_c, T\right)\varepsilon + h_1\left(T_c, t_c, T\right)\int_0^t \Delta C\left(\psi - \psi'\right)\frac{d\varepsilon}{d\tau}d\tau, \tag{6.1}$$

where

$$h_e\left(T_c, t_c, T\right) = \phi_0 - \frac{\omega^2}{a_2}\sum_m \frac{\left(A_m^0\right)^2}{B_m^0}; h_1\left(T_c, t_c, T\right) = \frac{\omega^2}{a_2}, \tag{6.2}$$

$$\Delta C(t) = \sum_m C^m \exp\left(-\frac{t}{\tau_m}\right); C^m = \frac{\left(A_m^0\right)^2}{B_m^0}, \tag{6.3}$$

and the reduced time variable function is

$$\psi(t) = \int_0^t \frac{d\zeta}{a\left(T_c, t_c, T\right)}d\zeta, \tag{6.4}$$

A_m^0 and B_m^0 are constants derived considering the first terms of the Helmholtz free energy expansion for a thermo-viscoelastic material, ω is the frequency and τ is reduced time. Omitting some lengthy derivations and assuming that the material is rheologically simple, implies that the temperature T enters only the reduced time expression, i.e. that (6.1) becomes

$$\sigma = h_e\left(T_c, t_c\right)\varepsilon + h_1\left(T_c, t_c\right)\int_0^t \Delta C\left(\psi - \psi'\right)\frac{d\varepsilon}{d\tau}d\tau. \tag{6.5}$$

For this type of materials the time-temperature shift in frequency domain is in the horizontal direction only. For polymers with changing degree of cure (characterized by cure temperature Tc and cure time tc), the simplest form of the shift function a in (6.4) is a product of the temperature shift function $a_T(T)$ and the cure shift function $a_c(T_c, t_c)$

$$a\left(T_c, t_c, T\right) = a_T(T)\,a_c\left(T_c, t_c\right). \tag{6.6}$$

If we additionally assume that *he* and h_1 do not depend on curing, the constitutive law simplifies to

$$\sigma = h_e \varepsilon + \int_0^t \Delta C(\psi - \psi') \frac{d\varepsilon}{d\tau} d\tau, \tag{6.7}$$

$$\psi(t) = \int_0^t \frac{d\zeta}{a_T(T) a_c(T_c, t_c)}. \tag{6.8}$$

In summary, in this paper we assess the assumption that only the reduced time depends on the curing parameters.

Two types of experiments were performed for this purpose. In the first type, we intend to assess the cure time (and hence consequently, degree of cure)/frequency superposition while in the second experiment we assess and characterise the temperature/frequency superposition for various degree of cure. It is in particular intended to verify the linearity of the viscoelastic response of the resin system in consideration. In the normal sense we talk about Time/Temperature superposition for response characterisation, in the present contribution however we examine the Time/Temperature/DoC superposition. However, as will be described later in the paper, it seems that using ToC (Time of Conversion) instead of DoC results in a linear relationship between time, temperature and ToC. Since the DoC (Degree of Cure) is a direct function of the ToC, we modify the present work by speaking in terms of Time/Temperature/ToC.

6.2 Method

6.2.1 Material

The resin system used for the characterisation of the thermo-viscoelasticity of a curing epoxy system consisted of LY5052 epoxy with HY5052 hardener from Huntsman. Prior to any experimental procedure, the resin was mixed in the ratio 100:38 by weight and stirred thoroughly, taking care not to include air bubbles. In total 18 samples was mixed and prepared for the DSC and 59 for the DMTA experiments.

6.2.2 Cure Kinetics

The cure kinetics characterisation was performed using a Diamond series DSC instrument from Perkin-Elmer. The samples for the DSC instrument were placed on a steel sample pans and weighed on a Mettler-Toledo AE240 digital scale. The cure kinetics profiling was performed for four different temperatures, 23, 40, 60 and 80 °C in two stages. The method adopted for 23 °C case was the following: the sample was firstly cured within the DSC for a predetermined time to obtain the residual heat of reaction and obtain the degree of cure. The predetermined times have been chosen as 2, 4, 6, 12 and 24 h. Three samples for each curing times were prepared and cured in the DSC at 23 °C. After each pre-curing stage a dynamic temperature scan from 0 to 200 °C, at the rate of 10 °C/min, was performed to obtain the residual heat of reaction. Another set of three samples containing uncured resins were put through the same dynamic temperature scan parameters as the case for 23 °C to obtain the average total heat of reaction for the system.

The second stage of DSC scanning was performed to obtain the heat of reaction for the rest of the temperatures, viz. 40, 60 and 80 °C. For this, isothermal scans were performed at the corresponding temperatures until equilibrium was attained. The sample was heated as quickly as possible to the corresponding temperatures so that no heat of reaction was lost during the heating up phase of the scan. The time required to attain equilibrium varies with temperature, with higher temperatures taking shorter times to attain equilibrium and vice-versa.

Epoxy resin curing is an exothermic reaction. Thermoset cure kinetics is typically modelled using *n*th-order or autocatalytic kinetic schemes in isothermal analysis [3, 11]. Resins obeying *n*th-order kinetics show a maximum reaction rate at $t=0$. In this work we have chosen to use a well established cure kinetic model by Kamal which describes the reaction rate as an analytical function of temperature and degree of cure as detailed in (6.9) [12],

$$\frac{d\alpha}{dt}(T,\alpha) = Ae^{-\frac{B}{RT}}(\alpha^m)(\alpha_{max} - \alpha)^n, \tag{6.9}$$

where α is the degree of cure, A is the pre-exponential rate constant (also called the collision frequency factor [13]), T is the temperature, B is the reaction energy, R is the ideal gas constant, m and n are constants independent of cure temperature. Moreover, the α is defined as

$$\alpha = 1 - \frac{H}{H_{tot}}, \tag{6.10}$$

where H is the residual heat of reaction and H_{tot} is the total heat of reaction. α_{max} is defined as the maximum degree of cure attained at equilibrium for various temperatures. The values for α_{max} are obtained for each cure temperature (T_{cure}) from the isothermal DSC experiments performed and fitted linearly to give the expression

$$\alpha_{\max} = 0.1433T_{cure} + 75.26. \tag{6.11}$$

6.2.3 Viscoelastic Characterisation

The thermo-viscoelastic characterisation of the resin system was performed using a TA Instruments Q800 DMA instrument. The choice of clamp for measurement was limited to a 3-point bending mode and single cantilever mode. Prior to the actual material characterisation, two different clamp geometries were evaluated for their characteristics. The procedure consisted of a frequency scan at various temperatures to obtain a master curve using fully cured neat resin samples of LY5052/HY5052. The scan was performed at a frequency range of 200, 100, 10, 1, 0.1 and 0.01 Hz and a temperature range of 23–200 °C in 10 °C increments. The test conditions are identical for both 3-point bend mode and single cantilever mode. In conclusion, we observed that the 3-point bend mode induces scatter in the results at higher temperatures. Therefore, the subsequent experiments were performed using single cantilever mode.

The experiments performed using the single cantilever mode required the use of uncured resin samples, consequently a method of manufacturing samples was devised such that samples were manufactured quickly and easily without an intermediate stage of making plates and cutting them down to size, which would be difficult considering that the samples are not fully cured. In order to efficiently prepare cantilever samples of uncured resin a moulds are made from a silicone rubber. The dimensions of the tool were designed so that the samples have dimensions of 17.5 mm × 12.8 mm × 3 mm. These dimensions are similar to the sample dimensions as recommended by TA Instruments for single cantilever mode sample 17.5 mm × 12.76 mm × 3.19 mm. The mould was then sandwiched between two plates of Teflon coated steel. The assembly was then clamped securely to prevent leakage once the mixed resin is poured into the mould. The mould assembly was then preheated to the required temperature in a convection oven before the resin is poured into. Once the resin was poured into the mould, it was left to cure for various predetermined times and temperatures to attain a certain degree of cure. After the predetermined curing time has been reached, the mould was removed from the oven and the samples are instantly frozen using a coolant spray to prevent additional curing and 'peeled' out from the mould. Three samples are produced at a time with identical degrees of cure with this method, although more could be made. The samples are then polished at the sides using a Struers Rotopol-1 polishing machine following which they are ready for DMTA testing. The samples were assessed visually to determine if they deformed or crack while being mounted on to the DMTA especially for samples that had low degrees of cure since they were most often in the transition or rubbery zones.

The DMTA experiments were performed at four different temperatures, identical to the DSC experiments, viz. 23, 40, 60 and 80 °C. The experiments were clustered into two groups to study the influence of degree of cure and the influence of temperature as well. In order to study the effect the degree of cure has on the viscoelastic properties, neat resin samples were cured at the aforementioned temperatures for various times and a frequency scan was performed on the samples. Table 6.1 shows the curing times at each of the four temperatures before testing.

Similarly to the DSC experiments, the DMTA experiments have been performed using two different approaches. In the first set, the samples were evaluated at a constant temperature in order to induce a predetermined degree of cure and observe how the modulus evolves with cure, while in the second set the samples were evaluated for the effect of varying temperatures. In the first set, the samples that are cured at a particular temperature were also tested at the same temperature. The frequency scan range for all the experiments are; 200, 100, 10, 1, 0.1 and 0.05 Hz, performed in a descending order. The second set of experiments was performed for 60 and 80 °C cure temperatures only. The curing times are identical to the previous experiments; however, the evaluated at different temperatures, viz. 23, 40, 60 and 80 °C. To be more precise, samples cured at 60 °C

Table 6.1 Curing temperature and times

Temperatures (°C)	Curing times (h)	Total number of samples
23	14.75; 17; 18; 19; 21; 24	6
40	3.5; 4; 4.5; 5; 5.5; 6; 6.5; 7; 7.5	9
60	1.25; 1.5; 1.75; 2: 2.25; 2.5; 3	28
80	0.5; 0.75; 1; 1.25	16

for 1.5 h were tested at 23, 40, 60 and 80 °C. Care was taken to ensure constant degree of cure during the experiments. This was by using one sample for each temperature. Since each of the DMA tests last approximately 10 min, the difference in degrees of cure from start to finish is negligible as confirmed by the DSC characterisation. The applied strain was in the order of 0.01 % in all DMTA tests.

6.3 Results and Analysis

6.3.1 DSC

The total heat of reaction, H_{tot} (cf. 6.10), for the presently considered resin system has been determined to be 482 J/g. The experimental results (data) from the DSC experiment are presented in Fig. 6.1 together with the fitted Kamal model. The fitting was performed using the data from the experiments performed at 40, 60 and 80 °C. The fitted parameters for the generic Kamal model (6.9) are presented in Table 6.2. Moreover, in Fig. 6.1 the Kamal model is assessed against the data from the 23 °C experiment. The model shows good agreement with the corresponding experimental data.

6.3.2 DMTA

The first set of experiments described in the previous section has been performed to determine the effects of the curing time on the viscoelastic behaviour at a constant temperature. In consequence, the raw data from each temperature has been used to construct master curves by shifting the data in the curing time domain. The reference curve that has been chosen for each of the temperatures was chosen such that the cure state is approximately between 80–85 % following which master curves are generated. In Fig. 6.2, each master cure is accompanied by the corresponding shift factors.

To determine how the master curves looks like in each case when an identical state of cure for each case is selected as a reference and put together. The reference cure state was chosen at 80 % and the curves were shifted to a value f as per the relation

$$f = m * \left(ToC - ToC_{Ref}\right) \tag{6.12}$$

Where

$$m = 0.0026T^2 - 0.0903T + 1.39$$

Here T is the cure temperature, ToC is the time of cure and ToC_{Ref} is the time of cure at which 80 % cure state for reference is attained. This has been shown in Fig. 6.3. The time of cure corresponding to 80 % for each temperature is shown in Table 6.3.

Following the shifting of the master curves into a position which represents 80 % state of cure, the individual master curves are shifted again in temperature, using 23 °C or room temperature as a reference state. This shifting in temperature is shown in Fig. 6.4. This inevitably requires the comparison of the shift factors with which the curves were shifted against the corresponding cure/testing temperature which are depicted in Fig. 6.5.

The next step is to develop and generate master curves for samples that have been cured to a certain degree of cure at a particular temperature and tested at various temperatures. The cure temperatures are 60 and 80 °C. The raw data correspond-

Fig. 6.1 Experimental DSC data and corresponding Kamal model fit

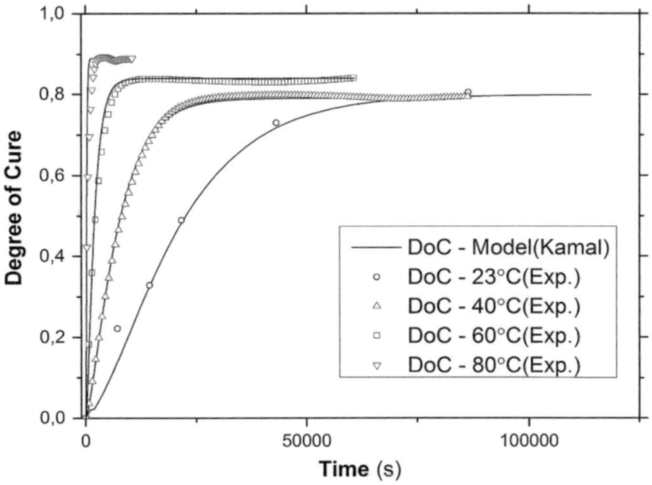

Table 6.2 Fitted parameters for the generic Kamal model

Parameter	Value
A[*1/s*]	274906.7
B[*J/mol*]	54888.35
m	0.25
n	1.01

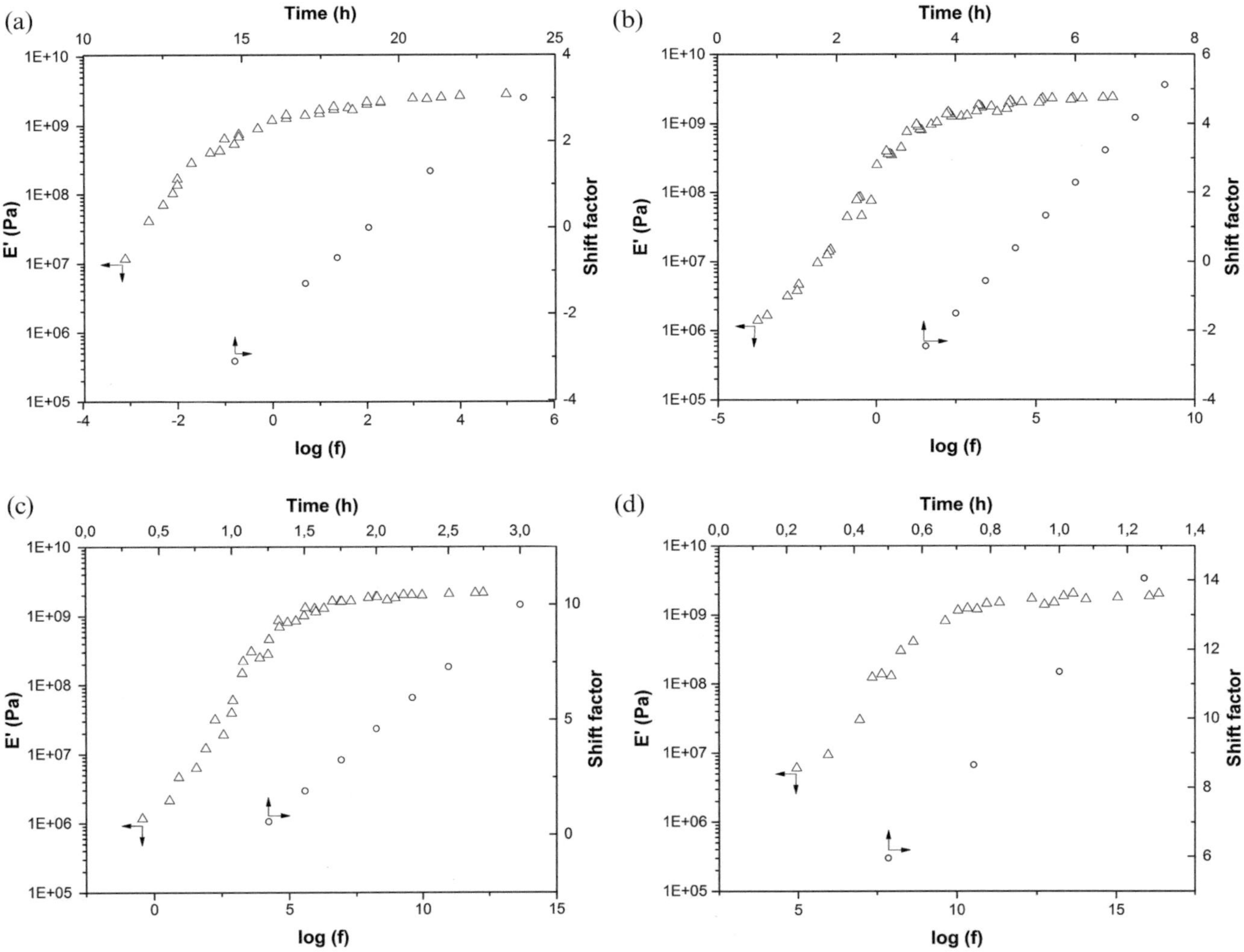

Fig. 6.2 Master curves by shifting in the curing time domain for (a) 23 °C, (b) 40 °C, (c) 60 °C and (d) 80 °C. The reference curve is shown in bold and is arbitrarily chosen to show the superposition. Note that it is the ToC/frequency domain superposition depicted here where the curves are shifted in cure. The shift factors used for the superposition are also (*arrows* indicate corresponding axes)

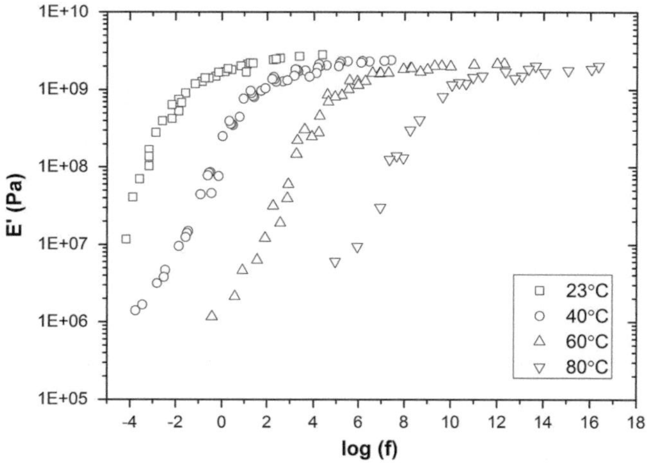

Fig. 6.3 Master curves for each temperatures with 80 % degree of cure as reference state

Table 6.3 Time required to attain 80 % DoC for each temperature based on the cure kinetic model

Temperature (°C)	Time to attain 80 % DoC (in minutes)
23	1221
40	288
60	68
80	19.5

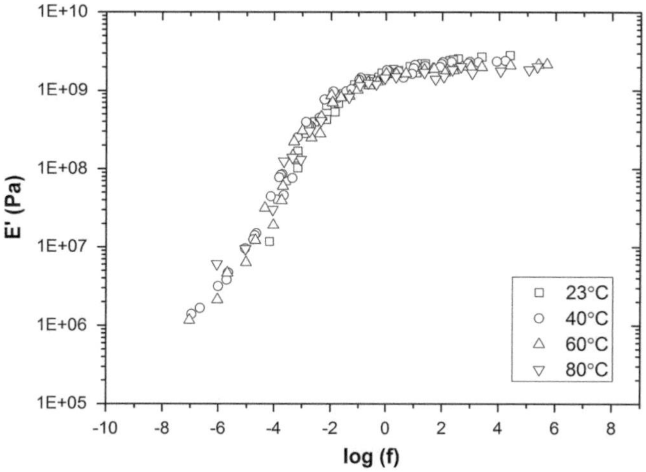

Fig. 6.4 Double shifted master curve combining data across all the cure temperatures

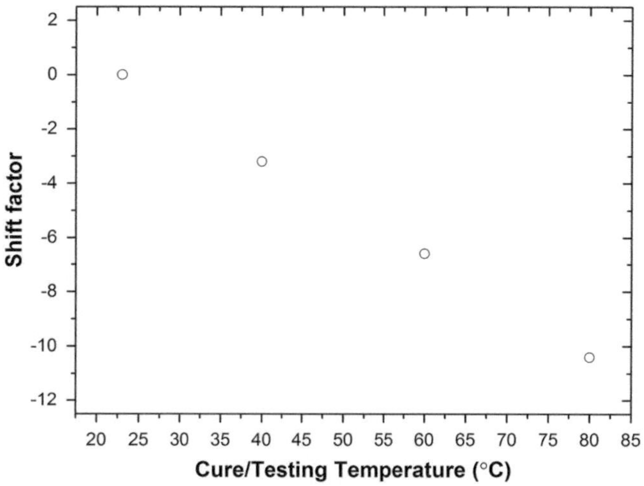

Fig. 6.5 Shift factors used to obtain the double shifted curve shown in Fig. 6.4 against the respective temperatures

ing to the frequency scans for epoxy cured at 60 °C for 1.5 h and 80 °C for 0.5 h is shown on Fig. 6.6 as an example. The curves thus generated were shifted and the individual master curves were combined for both 60 and 80 °C cases as shown in Fig. 6.7 for various cure times. The curve representing 23 °C testing temperature was chosen as the reference curve for the construction of master curves for all test cases.

The shift factors for each of the cases relating to both 60 and 80 °C experiments are constructed as shown in Fig. 6.8. The shift factors depicted in Fig. 6.8 show the shift in temperature for each of the individual master curves represented in Fig. 6.7.

Finally, as a last step the master curves obtained from the preceding operations for 60 and 80 °C were shifted once again in cure time. This double shifting was performed in ToC to obtain a single master curve. The reference curve for this operation was chosen at the time of cure corresponding to 80 % degree of cure as calculated in Table 6.3. The final shifted curves are represented in Fig. 6.9, along with the corresponding double shift factors for each ToC in Fig. 6.10.

6.4 Discussion and Concluding Remarks

In the first set of DMTA experiments, the samples have been cured at the same temperature to various degrees of cure. The characterisations were also performed at the same temperature at which they were cured. Four master curves, one for each curing temperature, have been constructed by shifting in cure time. These four master curves have been then used to

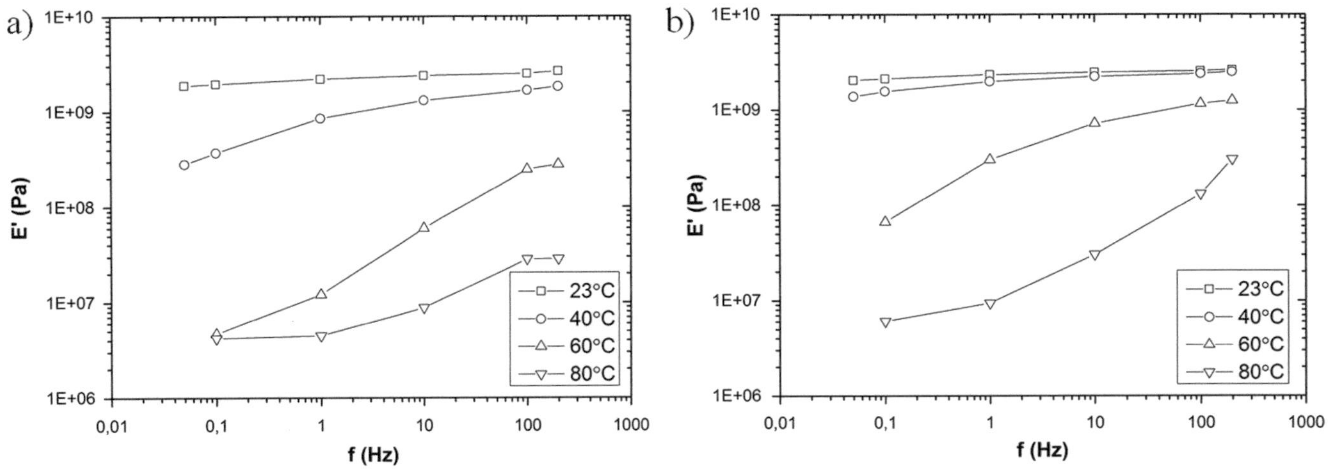

Fig. 6.6 Frequency scans for epoxy cured at (**a**) 60 °C for 1.5 h and (**b**) 80 °C for 0.5 h. The scans have been performed at various temperatures

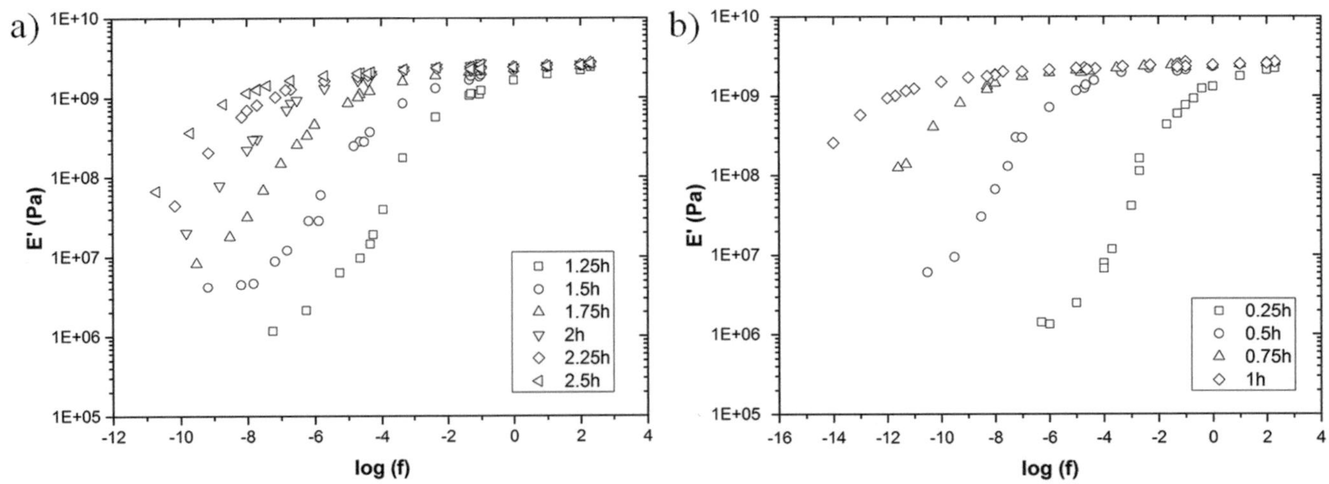

Fig. 6.7 Master curves for (**a**) 60 °C and (**b**) 80 °C relating to various degrees of cure. Note that the individual master curves generated for each cure state is shifted in temperature/frequency domain where the curves are shifted in temperature

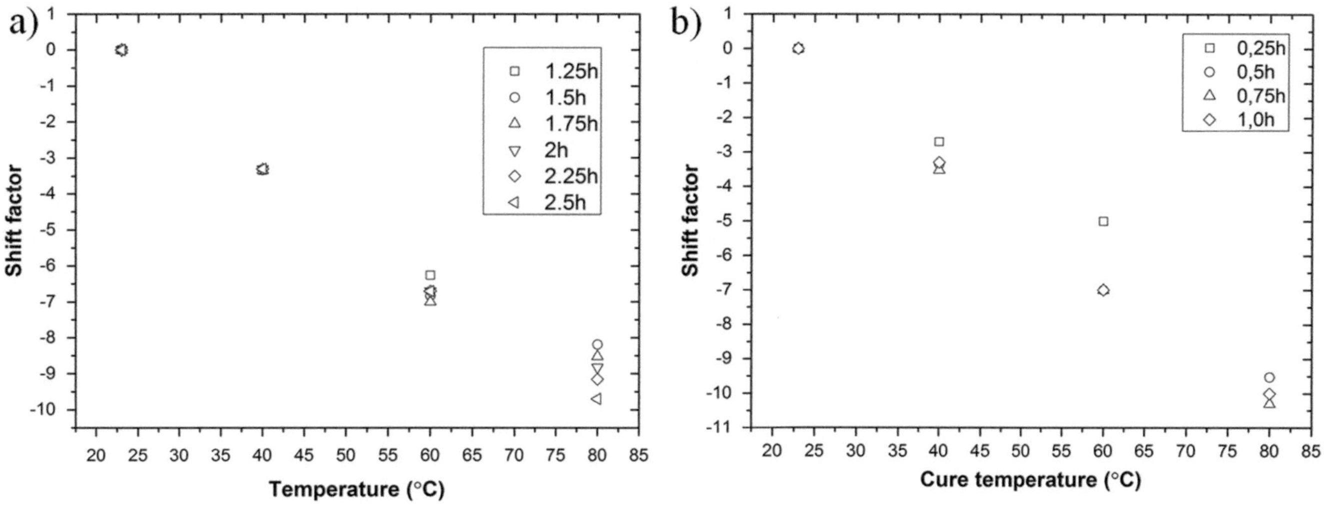

Fig. 6.8 Shift factors for (**a**) 60 °C and (**b**) 80 °C

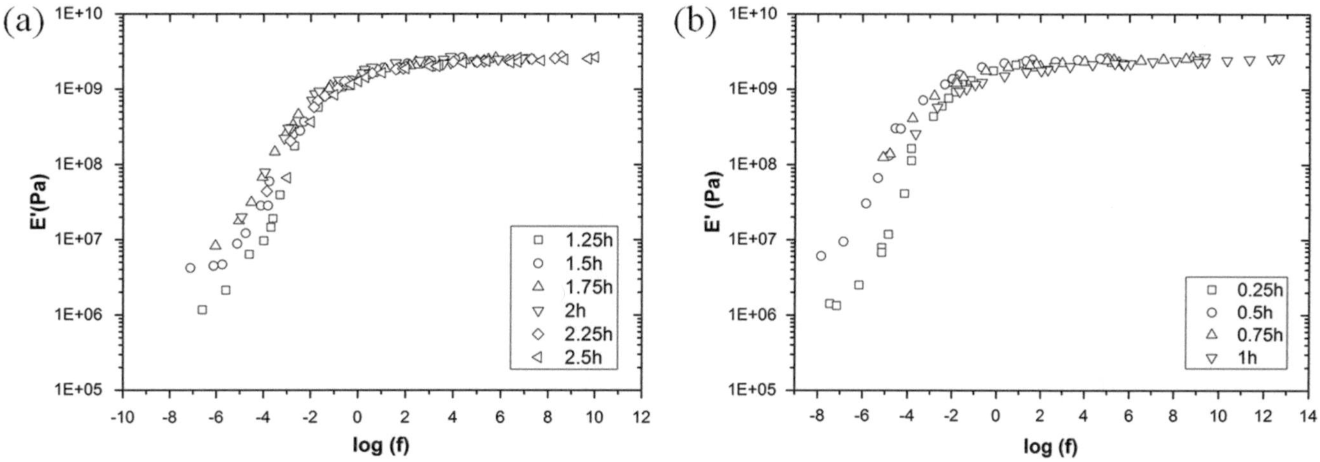

Fig. 6.9 Double shifted curves for (**a**) 60 °C and (**b**) 80 °C referenced at 80 % degree of cure in both instances

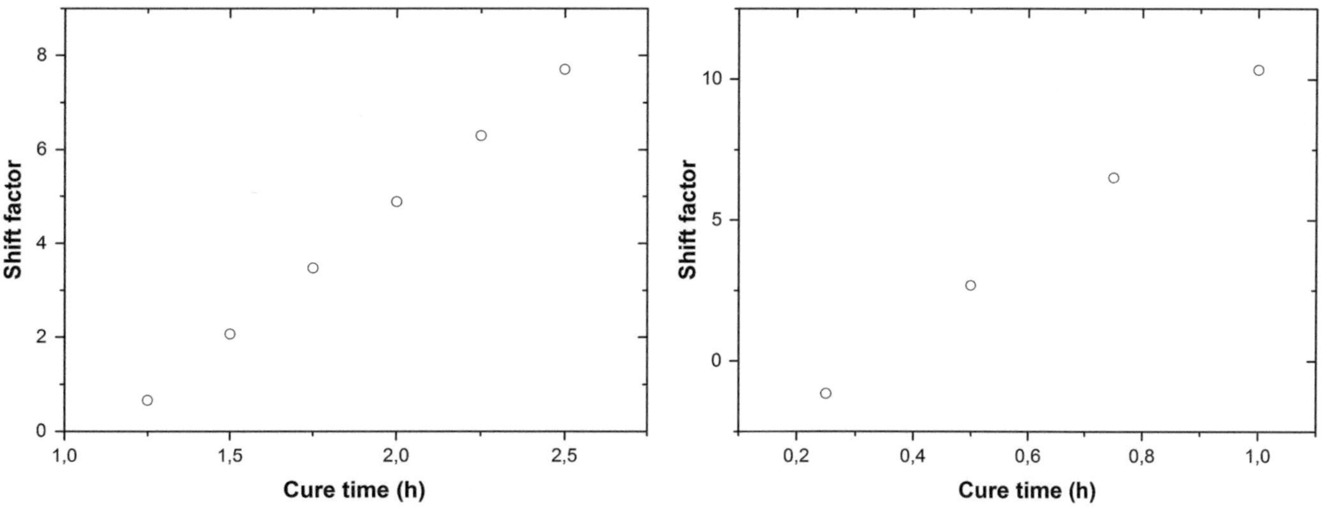

Fig. 6.10 Shift factors for the double shifted curves at (**a**) 60 °C and (**b**) 80 °C shown in Fig. 6.10

construct single 'super-master' curve by shifting them against temperature as shown in Fig. 6.4. In the second set of experiments, the samples were cured to identical degrees of cure using two different temperatures and then characterised at different temperatures. The master curves are then constructed by shifting temperature, and finally each individual master curves were shifted again, this time in cure time to obtain a 'super-master' curve as shown in Fig. 6.9.

We conclude from the first experimental set that the material behaviour is independent of the degree of cure since the shifts are linear with respect to both temperature and time. From the second experimental set, it seems that the master curves are independent of the degree of cure, which is true in the glassy region, as observed from Fig. 6.9. However, the rubbery plateau tends to shift vertically with degree of cure. The relationships for the shift factors for time and temperature are once again linear. The reason for the shifting rubber plateau is believed due to increased cross-link density. A striking result from the present work is that the final super-master curves, Figs. 6.4 and 6.9, and the corresponding shift factors, Figs. 6.5 and 6.8, are qualitatively equal, although being obtained using two different approaches. However, additional analysis needs to be performed to give a concretely prove the above statement. Finally, the above reasoning applies also to the hypothesis in equations (6.6–6.8), meaning that the shift factors can be assumed as a product of two independent shift factors dependent on temperature and degree of cure. Still, the analysis is only qualitative and needs additional examination.

References

1. Advani, G. Suresh, E. Murat Sozer, *Process Modeling in Composites Manufacturing* (CRC, Hoboken, 2012)
2. Y. Eom, L. Boogh, V. Michaud, P. Sunderland, J.A. Månson, Time-cure-temperature superposition for the prediction of instantaneous viscoelastic properties during cure. Polym. Eng. Sci. **40**(6), 1281–1292 (2000)
3. Y.K. Kim, S.R. White, Stress relaxation behavior of 3501-6 epoxy resin during cure. Polym. Eng. Sci. **36**(23), 2852–2862 (1996)
4. D.J. O'Brien, P.T. Mather, S.R. White, Viscoelastic properties of an epoxy resin during cure. J. Compos. Mater. **35**(10), 883–904 (2001)
5. R. Thorpe, A. Pousartip, Experimentally determining the viscoelastic behavior of a curing thermoset epoxy. 19th International Conference on Composite Materials, 1997–2002, (2013)
6. M. Sadeghinia, K.M. Jansen, L.J. Ernst, Characterization of the viscoelastic properties of an epoxy molding compound during cure. Microelectron. Reliab. **52**(8), 1711–1718 (2012)
7. K. Suzuki, Y. Miyano, T. Kunio, Change of viscoelastic properties of epoxy resin in the curing process. J. Appl. Polym. Sci. **21**(12), 3367–3379 (1977)
8. S.L. Simon, G.B. Mckenna, O. Sindt, Modeling the evolution of the dynamic mechanical properties of a commercial epoxy during cure after gelation. J. Appl. Polym. Sci. **76**(4), 495–508 (2000)
9. M. Zarrelli, A.A. Skordos, I.K. Partridge, Toward a constitutive model for cure-dependent modulus of a high temperature epoxy during the cure. Eur. Polym. J. **46**(8), 1705–1712 (2010)
10. D. Adolf, J.E. Martin, Time-cure superposition during crosslinking. Macromolecules **23**(15), 3700–3704 (1990)
11. R.B. Prime, E.A. Turi, Thermal characterization of polymeric materials. *Turi, EA, Ed*, 1380–1744 (1981)
12. J. Svanberg, Magnus, and J. Anders Holmberg. "Prediction of shape distortions Part I. FE-implementation of a path dependent constitutive model". Compos. A: Appl. Sci. Manuf. **35**(6), 711–721 (2004)
13. B. Bolasodun, O. Rufai, A. Nesbitt, R. Day, Comparison of the isothermal cure kinetics of araldite LY 5052/4 4'DDS epoxy system using a differential scanning calorimetry and a microwave heated calorimeter. Int. J. of Mater. Eng. **4**(4), 148–165 (2014)

Chapter 7
Structure–Property Relationships in Bimodal Polyethylene from Indentation Measurements

Aaron M. Forster, Ho Wei-Lun, Kar Tean Tan, and Don Hunston

Abstract Polyethylene is an advantageous material for the construction of buried pipelines. It is corrosion resistant, seismic tolerant, and utilizes low cost installation methods. The exceptional long term performance of polyethylene pipe has been fostered by an understanding of the structure–property relationships for the polyethylene chains, particularly the impact of molecular weight and short chain branching on tie chain formation. An area that remains a challenge for predicting performance is the thermal fusion bond created when two pipe sections are joined. The strength of the joint is predicated on the ability of the polyethylene chains to inter-diffuse and form inter-crystalline tie-chains across the two polyethylene surfaces. Testing the strength of the fusion bond is difficult because failure occurs under a biaxial stress state. A number of different destructive and non-destructive tests have been developed, but these allow ranking of joints. Instrumented indentation has the capability to measure elastic, plastic, and visco-elastic/plastic properties under a multi-axial stress state. In this work, instrumented indentation is used to develop structure–property relationships as a function of the local microstructure. Five different polyethylene resins used for pipe manufacturing are investigated. The impact of thermal processing is investigated by imposing three different thermal cooling histories (0.4, 9, and 100 °C/min) on the polyethylenes. The goal is to determine the impact of chain architecture, molecular weight, and crystallinity on modulus, hardness, yield stress, and viscoelastic properties. The results show that instrumented indentation using is capable of measuring many aspects of polyethylene behavior related to performance.

Keywords Indentation • Polyethylene • Yield stress • Hardness • Fusion bond

7.1 Introduction

Polyethylene (PE) is important for critical infrastructure applications such as drinking water, waste water, and fossil fuels transportation. The advantages of polyethylene compared to metallic pipe are competitive cost, lower corrosion rates, and toughness. Central to the durability of polyethylene is not only its low reactivity with the environment, but also the microstructure of the material. Slow crack growth (SCG) is the predominant long-time failure mode for these polyethylene materials. The process of SCG failure is understood as a stress concentrator, such as a dust, catalyst particle, or damage in the pipe causes small voids to form a crack in the pipe. These voids grow and form craze fibrils ahead of the stress concentration. The combination of loads and environmental exposure causes the stress concentrators to produce small voids that grow and form craze fibrils ahead of the stress concentration. The craze fibrils grow in length as the polyethylene chains disentangle from each other and the crystalline lamellae break up locally. The crack advances when the craze fibril breaks and stress shifts to the remaining fibrils, which causes them to break. Eventually the pipe fails in a brittle, rather than ductile, manner. This craze process is a creep driven failure and largely controlled by the tie-chains present in the pipe material and the SCB [1].

Early pipe technologies utilized a single molecular weight distribution with short chain branching (SCB) to reduce SCG failures. The advantage of SCB is that the branched molecules reside in the amorphous region between crystalline lamellae which makes disentanglement more difficult and increases the energy and time required to propagate a crack [2]. High performance resins used for piping applications are composed of a bimodal blend of two different polyethylene molecular masses: a medium to high density linear chain and a medium density branched chain (<10 branches per 1000 monomers).

A.M. Forster (✉) • H. Wei-Lun • D. Hunston
Engineering Laboratory, National Institute of Standards and Technology (NIST), 100 Bureau Dr., Gaithersburg, MD 20899, USA
e-mail: aaron.forster@nist.gov

K.T. Tan
Engineering Laboratory, National Institute of Standards and Technology (NIST), 100 Bureau Dr., Gaithersburg, MD 20899, USA

Department of Materials Science and Engineering, University of Maryland, College Park, MD 20742, USA

B. Antoun (ed.), *Challenges in Mechanics of Time Dependent Materials, Volume 2*,
Conference Proceedings of the Society for Experimental Mechanics Series, DOI 10.1007/978-3-319-22443-5_7

The advantage of the bimodal distribution is that the linear chains crystallize to provide the pipe its strength, while the branched chains resist chain pull-out from the amorphous regions giving the material exceptional slow crack growth resistance [2]. For example, Ting and coworkers have shown that bimodal formulations require higher force to initiate failure and have higher ultimate failure strains than unimodal resins [3].

Instrumented indentation is a measurement method that loads an axisymmetric indenter into a surface while measuring load and displacement with mN and nm resolution, respectively. The technique requires an understanding of the tip geometry, *area*=*f* (*contact depth*) and measures properties on μm³ volumes. Indenters may be pyramidal, conical, or hemispherical and they induce a multi-axial stress state from the combination of hydrostatic stress and shear stress. Indentation allows investigation of elastic (modulus), plastic (hardness), viscoelastic (creep), and viscoplastic (strain rate effects on hardness) properties of materials, which are related to the local microstructure and the effect on mechanical properties [4]. Indentation does not simulate the conditions near the crack tip during SCG crazing, but the geometry is similar to the opening of a crack under rock impingement.

The nuclear industry has been interested in the application of greater than 1 m diameter pipe for water transportation. While the structure–property relationships for pipe are generally understood, less is known about long term performance of thick walled pipe, particularly the fusion processing used to join these pipe sections. There has been concern that standard test methods for fusion joints do not adequately predict failure time for SCG [5]. In the past few years, the Welding Institute (TWI) in the UK has developed a whole pipe creep rupture test that is uniquely suited for long term failure measurements of fusion joints [6]. A challenge for any full pipe based test method is the cost required for evaluating large diameter pipe fusion joints. It would be advantageous to develop a coupon level metrology for conducting design of experiments and sensitivity analysis prior to testing large diameter pipes. Instrumented indentation is advantageous for investigating different areas of a fusion bond since the thermal and flow history in the joint region changes a local properties such that they differ from those of the parent pipe. The current research program at the National Institute of Standards and Technology (NIST) aims to utilized instrumented indentation to develop an understanding of the viscoelastic and viscoplastic behaviors for PE resin materials as a function of temperature, pressure, and cooling profiles observed during the formation of butt fusion joints in polyethylene. The initial focus is the heat affected zone (HAZ) around the fusion to provide an experimental route to maximize fusion bond conditions and provide the mechanical framework for modeling these thermal joints.

7.2 Experimental Details[1]

The five different PE resins (DOW) were measured for this work is shown in Table 7.1 The Pennsylvania Edge Notch Test (PENT) (ASTM F1473) is used in industry to determine the SCG resistance of polyethylene resins. PE-U is a medium density unimodal resin which has a PENT value of ~200 h and is classified as an ASTM 3408 resin with a hydrostatic design stress of 800 psi at 25 °C. PEb-1 through PEb-3 are bimodal resins with PENT times from greater than 5000 h to greater than 10,000 h as shown in Table 7.1. These are classified as ASTM 4710 resins with a hydrostatic design stress of 1000 psi at 25 °C. Peb-4 is a bimodal resin with a PENT time greater than 15,000 h. It is classified as an ASTM 2708 resin with a hydrostatic design stress of 800 psi at 25 °C.

Each resin was compression molded into a 15 cm × 15 cm × 500 μm plaque at 160 °C according to a procedure described in [7]. After pressing, the samples were removed from the press and placed in between two brass platens and manually clamped at all edges using C-clamps. The samples were reheated to 160 °C in an oven and cooled using either controlled cooling (0.4 and 8.7 °C/min) or quenched in an ice water bath (100 °C/min). The effect of thermal processing on the crystallinity, melting temperature, and lamellae thickness was characterized using Differential Scanning Calorimetry (DSC) scanning from room temperature to 160 °C at 10 °C/min. The lamellar thickness was calculated using the Gibbs-Thomson equation [8].

$$T_m = T_m^o \left(1 - \frac{2\sigma_e}{\Delta h_m l}\right) \tag{7.1}$$

where Tm is the measured melting temperature, Tm^0 is the equilibrium melting point of an infinite crystal (415 K), σe is the surface free energy of the basal plane (60.9×10^3 J/m²), Δhm is the enthalpy of fusion per unit volume (2.88×10^8 J/m³), and l is the lamellar crystal thickness.

[1] Certain commercial equipment and/or materials are identified in this report in order to adequately specify the experimental procedure. In no case does such identification imply recommendation or endorsement by the National Institute of Standards and Technology, nor does it imply that the equipment and/or materials used are necessarily the best available for the purpose.

Table 7.1 Properties of the five different polyethylene resins

		PE-U	PEb-1	PEb-2	PEb-3	PEb-4
		Unimodal	Bimodal PE	Bimodal PE	Bimodal PE	Bimodal PE
Density (g/cm^3)		0.946	0.946	0.949	0.949	0.941
Melt Index(g/10 min)	190 °C/21.6 kg	8.3	8.5	7.0	5.5	9.5
Tensile elongation (%)		850	>500	>500	>500	>600
Slow crack growth (h)		200	>5000	>10,000	>10,000	>15,000
Flexural modulus (psi)		120,000	150,000	150,000	150,000	>90,000
Tensile strength (psi)		3200	>3500	>3500	>3500	>2600

Bulk Testing: Bulk creep of selected resins was measured in tension using 10 mm × 10 mm rectangular bars. An initial stress of 2.75 MPa was applied to the bar at time equal to zero. The resulting change in strain from 0.17 to 0.51 % was measured over the 10,000 s. The creep compliance was calculated as a function of change in displacement with time. In order to determine whether the thermal treatment impacted the SCG resistance of the resins, SCG resistance was measured using a strain hardening (SH) technique [7, 8] conducted at 80 °C. To correlate with the indentation measurements, the strain hardening modulus was measured at room temperature. The strain hardening modulus was measured as the slope of the stress vs. strain curve, ($\langle G_p \rangle$), after the polyethylene natural draw ratio between an extension ratio of 8–12. The current study uses SH measurements of dog bone samples (ASTM D-638). The tests were performed at a constant displacement rate of 10 mm/min [7, 8] using an electromechanical test machine (MTS) and laser extensometer to measure strain. Prior to SH testing, the specimens were allowed to equilibrate at 80 °C in the oven for 30 min. A minimum of three tests were run for each type of specimen.

Indentation Testing: The creep behavior of the resins was measured using instrumented indentation with a 50 μm flat punch indenter over 600 s following the method of Herbert [10]. In this method a large step load ($P_1 = 80$ mN) is applied to the sample to achieve conformal contact with the flat punch. At this initial displacement, the sample is subjected to a stress-control step. The displacement (D_1) is held until the load reaches a plateau value (P_2). After the load ceases to change, an iterative control loop is initiated to measure the displacement change over a 30 s sampling time. The plateau value of the load (P_2) is held constant, while the displacement over time is measured for 30 s. The loop calculates whether the slope of displacement over time is below a drift threshold of 0.3 nm/s. If the material drifts beyond the threshold value, the controller moves the tip back to D_1 and holds the new load, P_3, for another iterative round. Once the drift threshold (0.3 nm/s) has been met the creep measurement may begin. This total time may be shorter than the 10×(*loading time*). In the creep measurement, the indenter applies a small load, P_4 and the displacement as a function of time is measured under this constant P_4. The creep under three different constant loads (P_4) of 3, 6, and 9 mN was investigated. In the time domain for instrumented indentation experiments the creep compliance using a flat punch may be calculated from (7.2) [10, 11].

$$J_c(t) = \frac{2Rh(t)}{P(1-\upsilon^2)} \tag{7.2}$$

where $Jc(t)$ is the creep compliance, R is the radius of the flat punch, $h(t)$ is the displacement time response, P is the applied constant load, and ν is the time independent Poisson's ratio (=0.35).

Indentation testing to measure modulus and hardness was conducted using a pyramidal (Berkovich) and conical indenter (10 μm 90°) tips. The tip area functions were determined by indenting fused silica (E = 72 GPa) to similar depths. The indentations were conducted using a continuous stiffness measurements (45 Hz, 2 nm) to a maximum depth of 1000 nm at a constant indentation strain rate. The strain rate was varied between 0.01 and 0.1 s^{-1}, but reported properties were determined at a strain rate of 0.05 s^{-1}.

7.3 Results and Discussions

The morphological characteristics within polyethylene are important for the ultimate properties of the material. The SCG behavior of polyethylene is a function of the molecular weight distribution, amount of short chain branching (SCB) and distribution of the short chain branches. Crystalline lamellae provide the stiffness, but these lamellae must be linked through the formation of chains that cross the amorphous region (tie chains) to provide strength. The amount and location of SCB affects the ability of the chains within the amorphous region to translate under stress which affects yield stress and SCG resistance [12]. The harder it is to disentangle those chains in the amorphous region, through the presence of short chain branching, the better the SCG resistance of the material. Slowing the speed of the material through the crystallization transition can increase the amount of order within the system, which may allow for the formation of larger crystals, less amorphous region, and reduce the tie chains. Quenching the material will decrease the order within the system, which will decrease the size of the crystalline lamellae, increase the amount of amorphous material, and potentially the tie chain density. Unfortunately, the formation of tie chains is not directly correlated to percent crystallinity (%X) or lamellar thickness [13, 14] measurements. Sardashanti [14] has shown that accelerating the cooling rate decreased the %X and decreased the lamellar thickness, which decreased the corrected strain hardening modulus for several polyethylenes.

Figure 7.1 shows the crystallinity for the polyethylene resins as a function of cooling rate. Increasing the cooling rate decreases the amount of crystallinity for Pe-U and Peb-4, which are expected to have lower Mw linear chains based on their lower hydrostatic design stress. Pe-U and Peb-4 had the greatest change in lamellar thickness going from 0.4 to 8.7 °C/min and were constant at the highest quench rate. This indicates these materials will exhibit a smaller fraction of crystallinity, but the crystal sizes will be more uniform during quenching. The higher strength materials, Peb-1 through Peb-3, exhibit a increase in crystallinity for slow cooling, but the crystallinity plateaus for the higher cooling rates. Peb-1 through Peb-3 showed a continual decrease in lamellar thickness with increasing cooling rate (not shown). This indicates that these materials will have a similar fraction of crystallinity, but the sizes of the crystals will be smaller with increasing cooling.

These morphological changes impact the performance of the material. In order to look at the impact on SCG resistance, the strain hardening modulus ($<G_p>$) at 80 °C was measured as a function of cooling rate, see Fig. 7.2. $<G_p>$ was 33 ± 3 MPa for the Pe-U and 48 ± 4 MPa for the Peb-1 through Peb-4 for the slow and medium cooling rates. These values agree with those found in the literature [7, 9, 13, 14] and show that this technique is able to measure differences between unimodal and bimodal materials that exhibit large differences in$<G_p>$ and PENT values. The strain hardening method was not able to discern differences between the Peb-1 through Peb-4 materials at the slow and medium cooling rates, which indicates the formation of tie chains was not affected by the cooling rate. These materials only exhibit a difference in PENT failure time of a factor of 3 compared to a difference factor of ~30 for the unimodal. Pe-U, Peb-3, and Peb-4 showed little sensitivity to

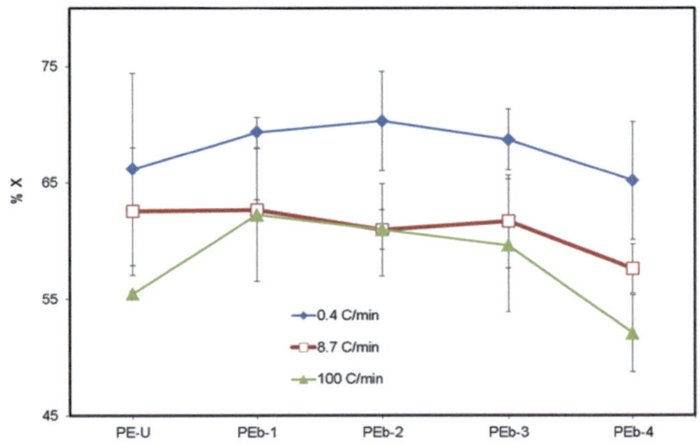

Fig. 7.1 Crystallinity of the polyethylene resins, measured from DSC, as a function of different cooling rates. The error bars represent the standard deviation about the mean of five measurements. When not visible, error bars are smaller than the data point

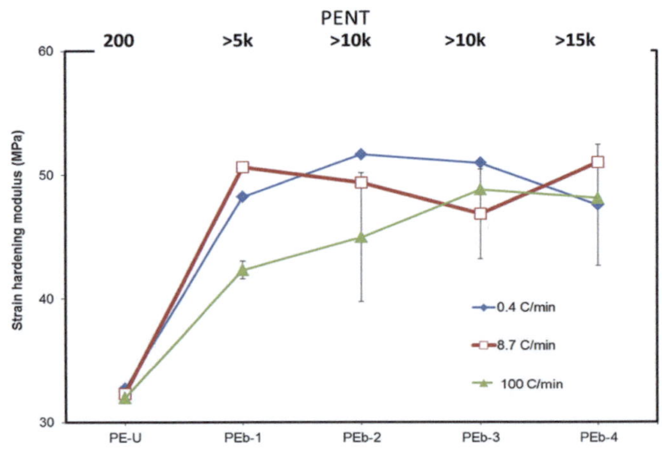

Fig. 7.2 Strain hardening modulus measured at 80 °C for the five different polyethylenes. The error bars represent the standard deviation about the mean of five measurements. When not visible, error bars are smaller than the data point

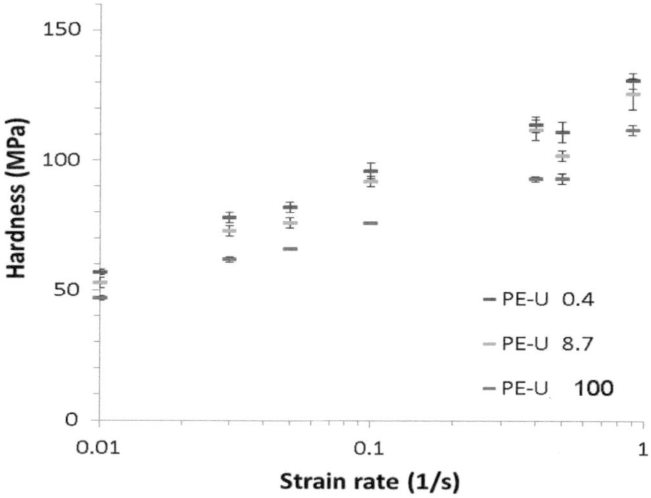

Fig. 7.3 Hardness as a function of strain rate for the Pe-U resin. The different points represent the different cooling rates and the error bars represent one standard deviation over 15 measurements

cooling rate which indicates that the changes in crystallinity and lamellar thickness did not significantly affect the formation of tie chains. Peb-1 and Peb-2 showed a lower $<G_p>$ for quench cooling, which indicates quenching reduces the ability of the material to form tie chains.

Hardness measurements have been shown to be sensitive to the crystallinity and yield stress within the sample [4]. Figure 7.3 shows the hardness for Pe-U as a function of the cooling rate. These measurements were conducted with the hemispherical tip. While the strain hardening modulus was not affected by the thermal treatment, the crystallinity was found to decrease with cooling rate. The higher crystallinity results in a higher hardness. Figure 7.3 shows that the hardness is linear with the logarithm of strain, which is an indication of an Eyring-type process (thermally activated), which has been demonstrated for polyethylenes. The process is similar for all cooling rates, but the dependence of hardness on strain rate increases with increasing crystallinity. Similar behavior has been observed for all of the other polyethylene materials tested. The highest strength bimodal materials, Peb-1 through Peb-3, exhibit the highest hardness and a similar dependence on the strain rate. Peb-4 has the lowest hardness values, but still exhibits the Eyring behavior with strain rate.

7.4 Conclusion

The impact of cooling rate on the microstructure, strain hardening, and creep behavior of five different polyethylene pipe materials was investigated. It was found that increasing the cooling rate affects the different resins in different ways. The high strength Peb-1 through Peb-3 materials was less sensitive to crystallinity changes at medium and quench cooling rates. Pe-U and Peb-4 exhibited a continual decrease in crystallinity with increasing cooling rate. The strain hardening behavior of the materials showed that a large difference in SCG resistance is needed to measure differences between the materials. In addition, only Peb-1 and Peb-2 were sensitive to the quench cooling rate. Instrumented indentation was able to measure the impact of cooling rate on hardness. For Pe-U increasing the cooling rate decreased the hardness, but the changes were non-monotonic. Hardness testing at different strain rates revealed a thermally activated process for plastic deformation under the indenter. This research shows that hardness measurements may not be sensitive to long term performance, such as slow crack growth resistance. It does show that hardness is an indicator of the crystal structure and a useful technique for probing the local microstructure of material within a fusion bond.

Acknowledgement Dr. Jimmy Zhou (DOW) kindly supplied the polyethylene resins for this study.

References

1. P. Davis, S. Burn, S. Gould, M. Cardy, G. Tjandraatmadja, P. Sadler, *Long-Term Performance Prediction for PE Pipes* (AWWA Research Foundation, Denver, 2007)
2. P.J. DesLauriers, D.C. Rohlfing, Macromol. Symp. **282**, 136–149 (2009)

3. S.K.M. Ting, J.G. Williams, A. Ivankovic, *Poly. Eng. Sci.*, 792–798 (2006)
4. F.J. Balta Calleja, Adv. Poly. Sci. **66**, 117–148 (1985)
5. E. Focht, NRC Issues regarding the use of HDPE piping in Safety-related Nuclear Applications; Use of HDPE for Power Plant Piping Systems Workshop, Charlotte, NC, 7–10 June 2010
6. M. Troughton, C. Brown, J. Hessel, M. Piovano, Proceedings of the Plastics Pipes XIII Conference, Washington, DC, October 2006
7. L. Kurelec, M. Teeuwen, H. Schoffeleers, R. Deblieck, Polymer **46**, 6369–6379 (2005)
8. A. Wlochowicz, M. Eder, Polymer **9**, 1268–1270 (1984)
9. M. McCarthy, R. Deblieck, P. Mindermann, R. Kloth, L. Kurelec, H. Martens, Plastic Pipes XIV, Budapest, Hungary, 22–24 Sept 2008
10. E.G. Herbert, W.C. Oliver, A. Lumsdaine, G.M. Pharr, J. Mater. Res. **24**, 626–637 (2009)
11. H. Lu, B. Wang, J. Ma, G. Huang, H. Viswanathan, Mech. Time-Dep. Mat. **7**, 189–207 (2003)
12. Y.-L. Huang, N. Brown, J. Poly, Sci. B Poly. Phys. **29**, 29–137 (1991)
13. J. Cheng, M. Polak, A. Penlidis, J. Macromolecular Sci, Part A Pure Appl. Chem. **45**, 599–611 (2009)
14. P. Sardashti, A.J. Scott, C. Tzoganakis, M. Polak, A. Penlidis, J. Macromolecular Sci, Part A Pure Appl. Chem. **51**, 189–202 (2014)

Chapter 8
Prediction Validation of Thermal Aging Performance of Military Composite Bridges

Frank Abdi, Saber DorMohammadi, and Mohamad Reza Talagani

Abstract Short-term aging properties of a tri-axial polymer composite material for mobile bridges operating in high heat environment were analyzed with progressive failure simulation software and physically tested for verification. The work focused on thermal cycling, impact damage and ultra violet (UV) exposure effects on degradation of the composite material tensile and compressive strengths as well as fatigue life. Tests and simulations showed that short-term thermal cycling and UV exposure did not have significant effects on the composite mechanical properties. Simulation results agreed well with the test data.

Keywords Aging • Creep • Degradation • Composite architecture • Curing and distortion • Residual stress • Material modeling

8.1 Introduction

Exposure of advanced polymeric matrix composite materials (PMC), such as graphite/epoxy, to "real life" use/environment may result in irreversible changes in the original properties of the material. Close attention to design details and using the best design/manufacturing methods helps retardation but does not eliminate it. This process of change over time is loosely referred to as "aging". Material aging may translate to structural changes in mission-critical components that can have potentially catastrophic effects. In this study, the environmental thermal aging failure of the polymer composite material employed in a military mobile bridge was investigated using both test and simulation methods. The commercial software, GENOA, was utilized in the environmental aging analysis of the composite laminate. The experimental tests were conducted at the Orange County Materials (OCM) Test Lab. The effort focused on the aging characterization of the bridge composite constituents and identification of the environmental effect on the laminate mechanical performance. Since advanced composites only recently have been exploited for bridge research and development (R&D) studies, this work provides useful analytical tools for composite bridge design including material long/short term environmental aging effects.

8.2 Composite System

The composite system used in the test panels was tri-axial carbon fiber reinforced Epon, [45/−45/90]s. The mechanical loading direction was in the 0° direction. The static properties of the constituents were previously derived from numerical material characterization of the Composite Army Bridge (CAB) system [1] and adopted for the current study of the tri-axial composite laminate. The constituent static properties used were:

Carbon Fiber:
Modulus, Ef = 37msi
Coefficient of thermal expansion, CTEf11 = −4.300E-07 in/in/F; CTEf22 = 5.6000E-06 in/in/F
Strength, Sf_tension = 500 ksi; Sf_compression = 350 ksi

F. Abdi (✉) • S. DorMohammadi • M.R. Talagani
AlphaSTAR Corporation, 5150 E. Pacific coast highway Suite. 650, Long Beach, CA 90804, USA
e-mail: fabdi@alphastarcorp.com

B. Antoun (ed.), *Challenges in Mechanics of Time Dependent Materials, Volume 2*,
Conference Proceedings of the Society for Experimental Mechanics Series, DOI 10.1007/978-3-319-22443-5_8

Epon:
Modulus, Em = 0.45 msi
Poisson ratio, Num = 0.4
Coefficient of thermal expansion, CTEm 0.2700E-04 in/in/F
Strength, Sm_tension = 3.8 ksi; Sm_compression = 16ksi; Sm_shear = 6.5 ksi

The composite tension and compression properties simulated using these constituent properties agreed well with the CAB test data [1] as shown in Table 8.1.

The fatigue (S–N) degradation curve for the matrix at a stress ratio R = −1 (Fig. 8.1) was derived from the material calibration process based on test data. Published ultra-violet degradation for epoxy (Fig. 8.2) [2], was adopted for this study, which might not reflect accurate UV degradation for the tri-axial material in this study.

Table 8.1 Comparison between GENOA simulated tri-axial composite static properties and the CAB test data

	Previous CAB test [1]		GENOA simulation	
Property	Modulus (msi)	Strength (ksi)	Modulus (msi)	Strength (ksi)
Tension	7.8	101.0	8.2	98.0
Compression	7.8	75.8	8.2	73

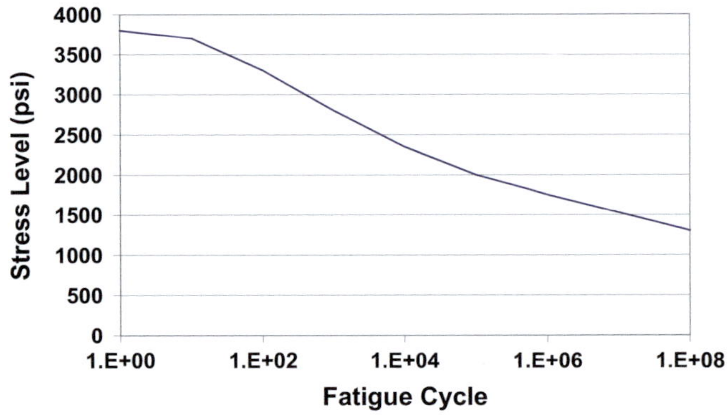

Fig. 8.1 Fatigue S-N degradation of the matrix

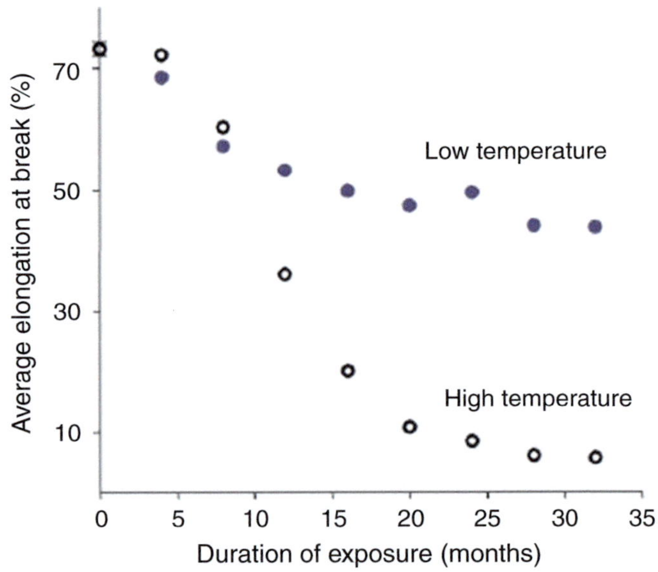

Fig. 8.2 UV degradation of Epon [2]

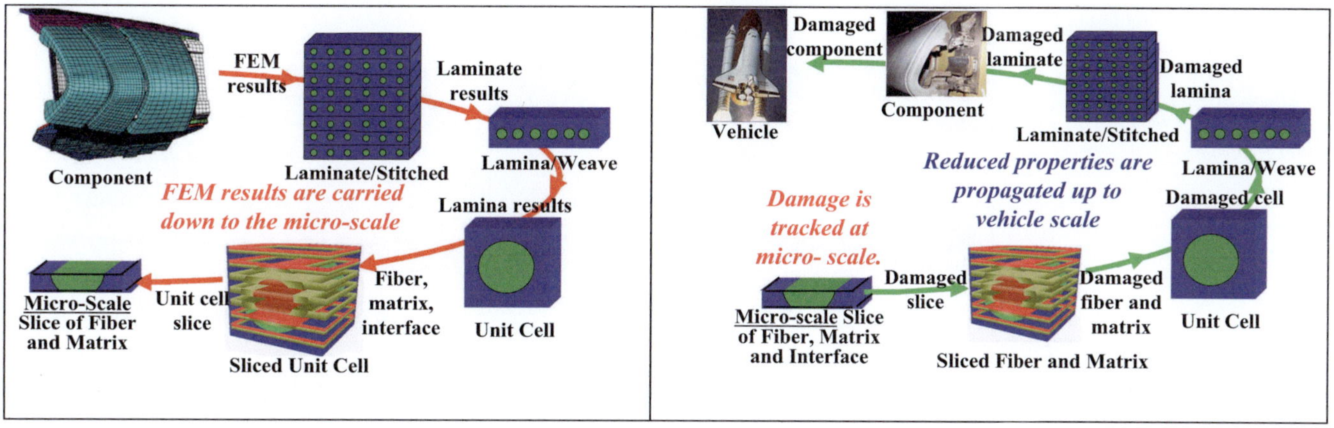

Fig. 8.3 The GENOA micro–macro progressive failure analysis process

8.3 Analysis Method

The GENOA software uses micro and macro interaction methods in the composite structural progressive failure analysis procedures (Fig. 8.3) [3]. Micro stress and damages are computed on the constituent level and the corresponding material degradation is reflected in the macroscopic finite element structural stiffness. In this study, the simulation procedure followed the test procedure and the damage occurring in each step was transmitted to the subsequent process.

8.4 Test and Simulation Procedures

The effects of thermal cycling and ultra-violet exposure on the mechanical properties of the tri-axial composite material were investigated using both physical testing and numerical simulation. Details of the testing and simulation are described below.

- Control—Mechanical properties without thermal cycling or UV exposure
  Three mechanical properties of the laminate were tested and simulated (Fig. 8.4a):
  1. Static tension
  2. Static compression
  3. Tension/compression fatigue (R = −1)
- Effects of no-load thermal cycling (50 cycles from −27.4 to 159.8 °F)
  There were four tests in this group (Fig. 8.4b):
  1. No-load thermal cycling
  2. Static tension after no-load thermal cycling
  3. Static compression after no-load thermal cycling
  4. Tension/compression fatigue (R = −1) after no-load thermal cycling.
  5. Impact and post-impact mechanical properties after no-load thermal cycling (50 cycles from −27.4 to 159.8 °F)
- The test specimens underwent impact after the no-load cycling process. Subsequent to the impact, tension, compression and fatigue tests were performed to study the post-impact residual properties of the tri-axial composite material (Fig. 8.4c):
  1. No load thermal cycling
  2. Impact after no-load thermal cycling
  3. Tension subsequent to the impact
  4. Compression subsequent to the impact
  5. Tension/compression fatigue (R = −1) subsequent to the impact.

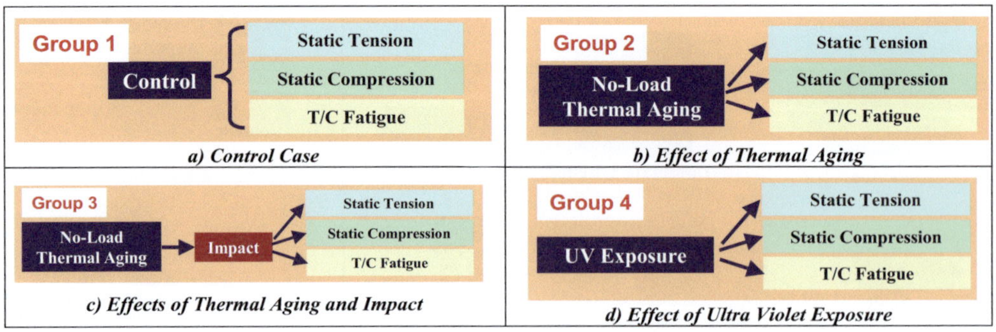

Fig. 8.4 Test and simulation procedures. (**a**) *Control Case*. (**b**) *Effect of Thermal Aging*. (**c**) *Effects of Thermal Aging and Impact*. (**d**) *Effect of Ultra Violet Exposure*

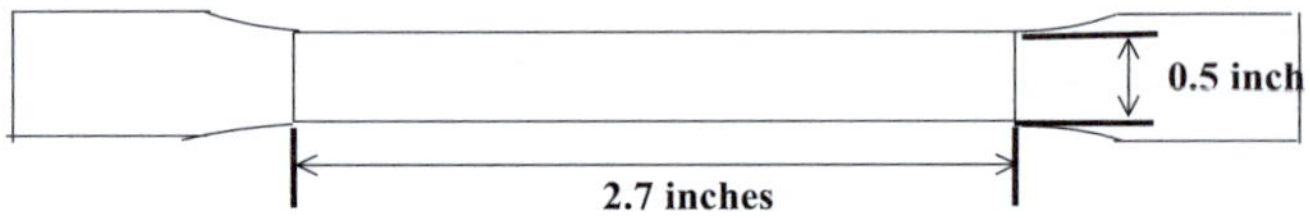

Fig. 8.5 Tension specimen (ASTM D 638-03)

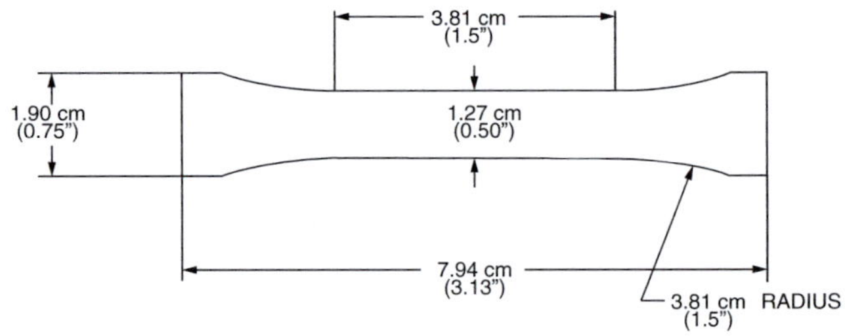

Fig. 8.6 Compression Specimen (ASTM D 695-02a)

- Effect of ultra-violet exposure
  This test group consisted of four tests (Fig. 8.4d):

  1. Ultra-violet process
  2. Tension subsequent to the UV exposure
  3. Compression subsequent to the UV exposure
  4. Tension/compression fatigue (R = −1) subsequent to the UV exposure.

8.5 Test Specimens

Specimen dimensions for tension and compression tests are shown in Figs. 8.5 and 8.6 Specimens were fabricated per ASTM standards D638-03 and D695-02a, respectively. The fatigue test specimen was a half-inch wide. T The panel dimensions of the impact test were 13 × 7 in. (Fig. 8.7). Ten impact tests were conducted on the panel. The area fixed on the impact machine for each impact was 3 × 5 in.

8.6 Simulation Models

The tension, compression, and fatigue specimens after thermal cycling or UV exposure were modeled using 200 shell elements (Fig. 8.8). The ASTM test specifications for these three specimens were not critical for simulation as long as a uniform stress state could be achieved within the specimen gauge length. Figure 8.9 illustrates the finite element model for the impact simulation, in which the area enclosed by the dash-line box represents the effective area (3 × 5 in.) fixed on the impact machine. The post-impact tension, compression, and fatigue model were directly extracted from the impact model to inherit the impact damage as shown in Fig. 8.9. The entire impact model, including the impactor, comprised 1708 shell elements.

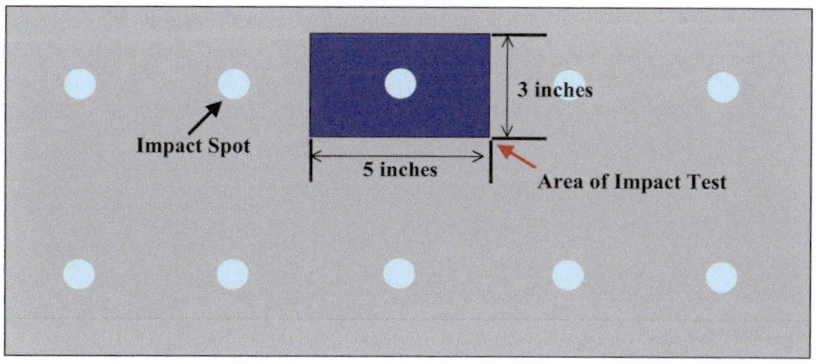

Fig. 8.7 Panel of drop impact test. Ten spots were impacted and the area set up on the impact machine for each impact is 3×5 in.

Fig. 8.8 Finite element model of tension, compression or fatigue specimens after thermal cycling or UV. Only the gauge length was modeled

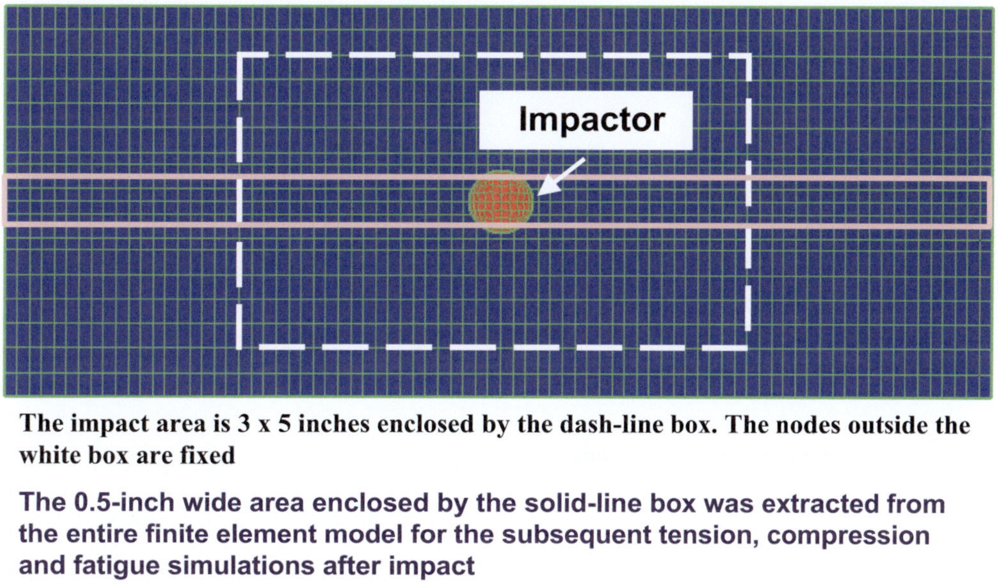

Fig. 8.9 Finite element model of impact specimen. The 0.5-in. wide model for post-impact tension, compression and fatigue simulations was extracted from the entire impact model to inherit the impact damage

8.7 Test and Simulation Results

8.7.1 Control: Mechanical Properties Prior to Thermal Cycling or UV Exposure

The tension, compression and fatigue tests of the tri-axial composite were conducted prior to thermal cycling or UV exposure for the baseline study of this material. Table 8.2 shows the comparison between the OCM tested and GENOA simulated results. GENOA simulation for this material was performed using the constituent properties derived from the CAB test data [1], which caused some deviation between the tested and simulated composite moduli for the current composite material.

The comparison between OCM tested and GENOA simulated composite fatigue life, prior to thermal cycling or UV exposure is shown in Fig. 8.10. The fatigue test and simulation were both performed at various stress levels with stress ratio R=−1. The fatigue life after a million cycles is defined as the run out condition. The simulation result was consistent with the test data.

Table 8.2 Comparison Between OCM Tested and GENOA Simulated Composite Static Properties

	OCM test		GENOA simulation	
Property	Modulus (msi)	Strength (ksi)	Modulus (msi)	Strength (ksi)
Tension	10.4	94.4	8.2	98.0
Compression	10.3	74.0	8.2	73

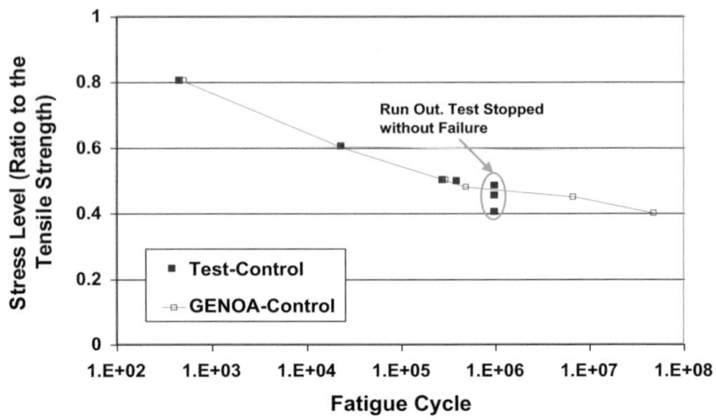

Fig. 8.10 Comparison between OCM-tested and GENOA-simulated composite fatigue lives in the baseline study prior to thermal cycling or UV exposure

8.7.2 *No-Load Thermal Cycling and Subsequent Mechanical Loading*

No-Load Thermal Cycling (50 Cycles from −27.4 to 159.8 °F)

No-load thermal cycling was conducted by varying the environmental temperature of the specimen from −27.4 to 159.8 °F for 50 cycles without applying any load. GENOA simulation shows there was no damage occurring during this process.

Tension Subsequent to the No-Load Thermal Cycling

Tensile properties of the tri-axial composite were tested after 50 no-load thermal cycles and were compared with simulation results (Figs. 8.11 and 8.12). Simulation results did not detect any obvious differences between the properties of the control and no-load thermal cycling cases because no damage occurred during the simulated 50 thermal fatigue cycles. In addition, the effect of thermal cycling on the composite modulus was opposite to its effect on the composite tensile strength in the test data. Therefore, it is reasonable to say that the 50 no-load thermal cycles did not affect the tri-axial composite tensile properties.

Compression Subsequent to the No-Load Thermal Cycling

The compressive properties of the tri-axial composite were tested after 50 no-load thermal cycles and were compared with simulation results (Figs. 8.13 and 8.14). Similar to the tensile case, simulation results did not detect differences between the properties of the control and no-load thermal cycling cases. Additionally, the effect of thermal cycling on the composite modulus was opposite to its effect on the composite compressive strength in the test data. Therefore, it is safe to say that the 50 no-load thermal cycles do not affect the tri-axial composite compressive properties.

Mechanical Fatigue Subsequent to the No-Load Thermal Cycling

The fatigue life of the tri-axial composite was tested at various stress levels after 50 no-load thermal cycles. The mechanical fatigue stress ratio of each cycle was, $R=-1$. The tested and simulated results are shown in Fig. 8.15. Similar to the tensile and compressive cases, the 50 no-load thermal cycles did not have an obvious effect on the composite fatigue life.

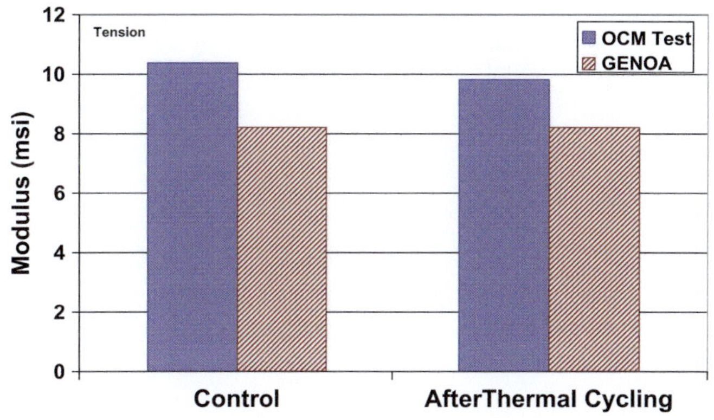

Fig. 8.11 Composite tensile moduli after 50 no load thermal cycles

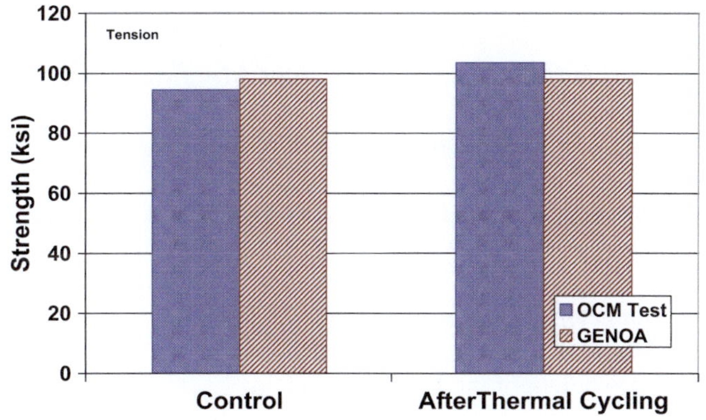

Fig. 8.12 Composite tensile strengths after 50 no load thermal cycles

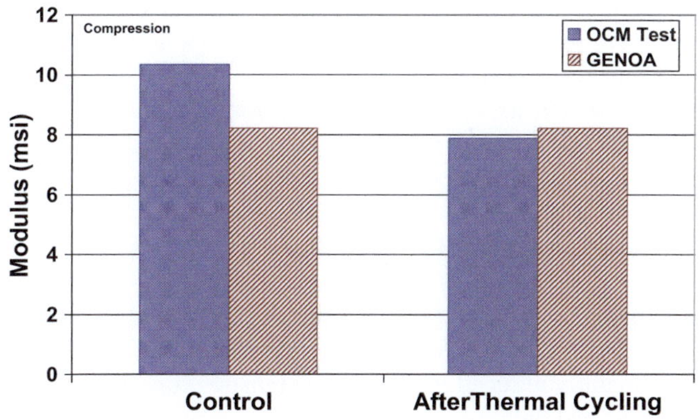

Fig. 8.13 Composite compressive moduli after 50 no load thermal cycles

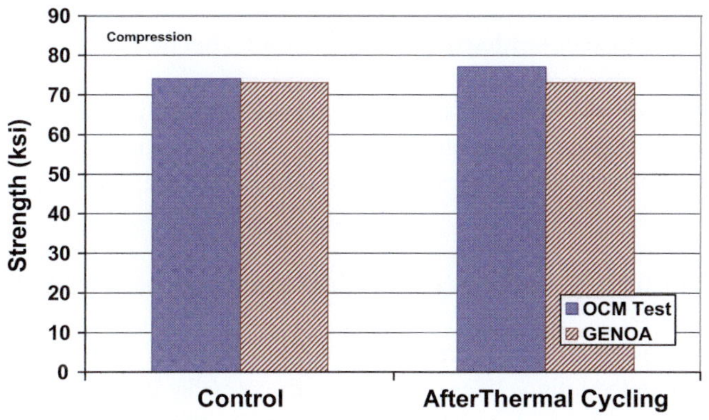

Fig. 8.14 Composite compressive strengths after 50 no load thermal cycles

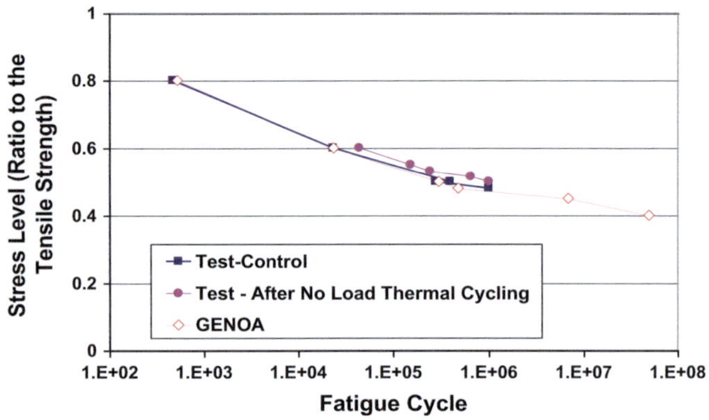

Fig. 8.15 Composite fatigue lives prior to and after 50 no load thermal cycles

8.7.3 Impact and Post Impact Mechanical Properties After 50 No-Load Thermal Cycles

Impact After No-Load Thermal Cycling

A drop impact test was conducted on the tri-axial composite laminate after 50 no-load thermal cycles. The impact test parameters were:

- Diameter of the impactor head: 0.625 in.
- Weight of the impactor: 10.8 lbs
- Drop height of the impactor: 30.8 in.

The composite panel in the impact test and simulation is shown in Figs. 8.7 and 8.9. GENOA-LSDYNA 3D was employed in simulation. LSDYNA 3D calculated the explicit dynamic impact process while GENOA dealt with the composite micro-stress computation, damage/failure detection, and material degradation due to damage.

Figure 8.16 compares the tested and simulated damage areas in the panel after impact. The test picture is a C-Scan graph (2¼ MHZ. Ultrasonic Thru-Transmission C-Scan). It can be seen that the simulated impact damage distribution agrees reasonably well with the test result.

Tension Subsequent to the No-Load Thermal Cycling and Impact

The panel after impact was cut into a 0.5-in. wide specimen (Fig. 8.9) for tension, compression or fatigue test. Figures 8.17 and 8.18 show the post impact tensile properties of the tri-axial composite laminate. It can be seen that the impact damage significantly reduced the composite tensile properties. Simulated composite tensile modulus and strength after impact agree with the test data.

Compression Subsequent to No-Load Thermal Cycling and Impact

Post-impact compressive modulus and strength of the tri-axial composite laminate were tested. Results are plotted in Figs. 8.19 and 8.20 along with simulation results for comparison. Similar to the post-impact tension case, the composite compressive properties were also reduced by the impact damage. The simulation results agree well with the test data.

Mechanical Fatigue (R = −1) Subsequent to the No-Load Thermal Cycling and Impact

A tension-compression fatigue load (R = −1) was applied to the specimens cut from the impacted laminate panel. Tested and simulated fatigue lives of the tri-axial composite laminate are shown in Fig. 8.21. The impact damage shown in Fig. 8.16 significantly reduced the composite fatigue life as illustrated in Fig. 8.21. The simulation results are consistent with the test data.

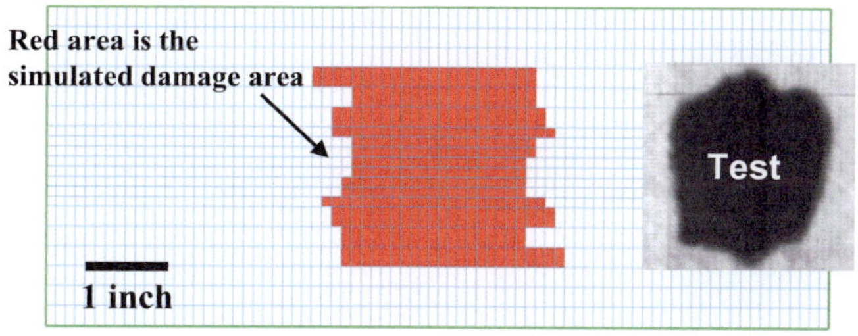

Fig. 8.16 Comparison between simulated and tested damage areas in the laminate after impact

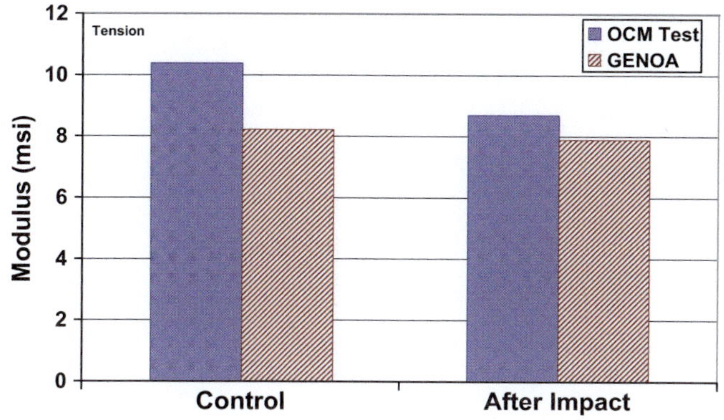

Fig. 8.17 Comparison between composite tensile moduli prior to and after impact

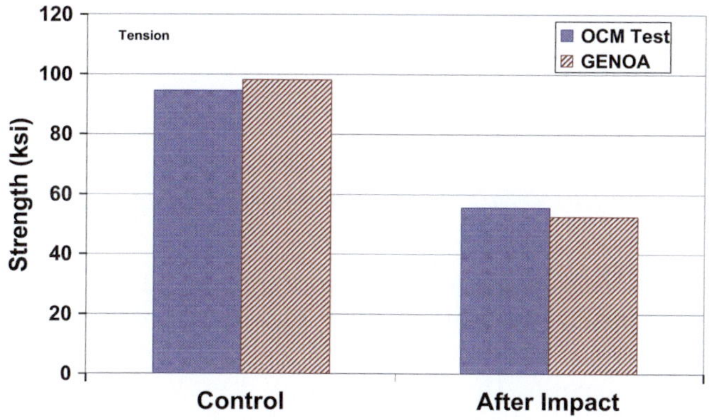

Fig. 8.18 Comparison between composite tensile strengths prior to and after impact

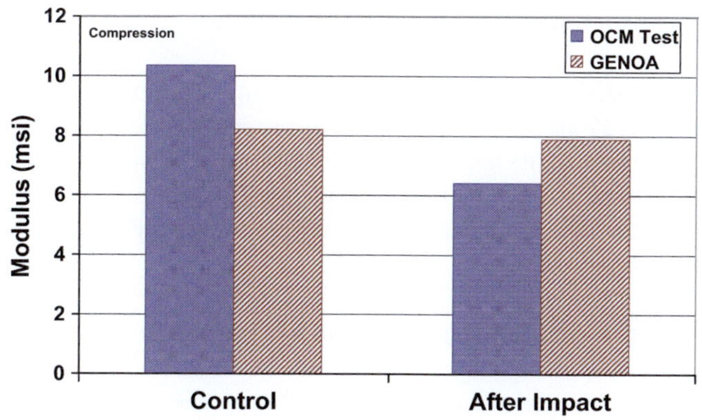

Fig. 8.19 Comparison between composite compressive moduli prior to and after impact

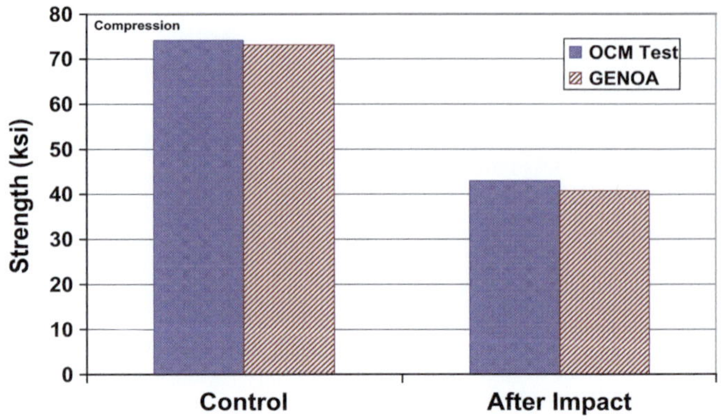

Fig. 8.20 Comparison between composite compressive strengths prior to and after impact

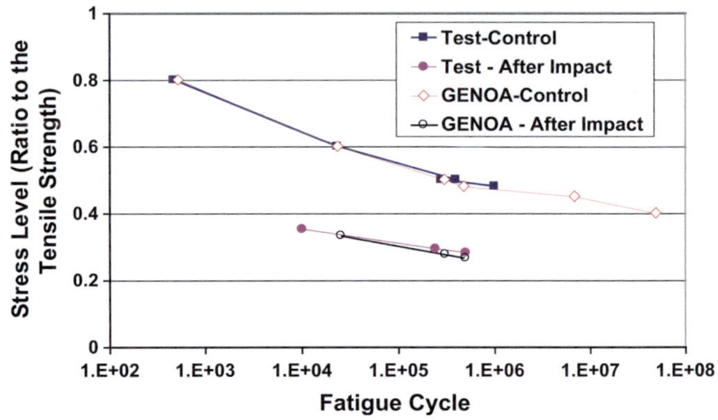

Fig. 8.21 Comparison between composite fatigue lives prior to and after impact

8.7.4 Subsequent Mechanical Loading After UV Exposure

The tri-axial composite specimens were exposed to 313-bulb ultra-violet for 750 h. The specimens were then tested for tension, compression and fatigue properties. According to Fig. 8.2, epoxy degradation at 750-h UV exposure is less than 1 % [2]. It is possible that the UV exposure in the OCM test did not create obvious degradation in the matrix (Epon) of the tri-axial composite. The property and UV degradation relationship defined in Fig. 8.2 were adopted for the GENOA simulation since UV degradation data for Epon was not available.

Tension Subsequent to the UV Exposure

The tested and simulated tensile moduli and strengths of the tri-axial composite after 750-h UV exposure are shown in Figs. 8.22 and 8.23. The composite properties prior to UV exposure are also presented for comparison. The reductions of the tested composite tensile modulus and strength after UV exposure might not be caused by the UV effect because the composite compressive properties and fatigue life were not consistently reduced as shown in the following sections. Therefore, it is reasonable to interpret the property reduction in Figs. 8.22 and 8.23 as material variation.

Compression Subsequent to the UV Exposure

The tested and simulated compressive properties of the tri-axial composite after 750-h UV exposure are shown in Figs. 8.24 and 8.25. The composite compressive properties prior to UV exposure are also illustrated. The composite compressive modulus after UV exposure is a little lower but the compressive strength is slightly higher compared to the control case. Due to the opposite variation trends of the composite modulus and strength, it is safe to say that the variations were not caused by the UV exposure but by the material variability. The simulation results also show that the effect of 750-h UV exposure was negligible.

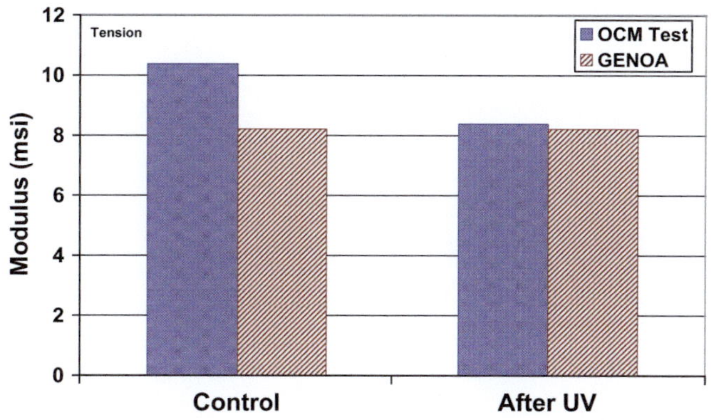

Fig. 8.22 Composite tensile moduli after 750-h UV exposure

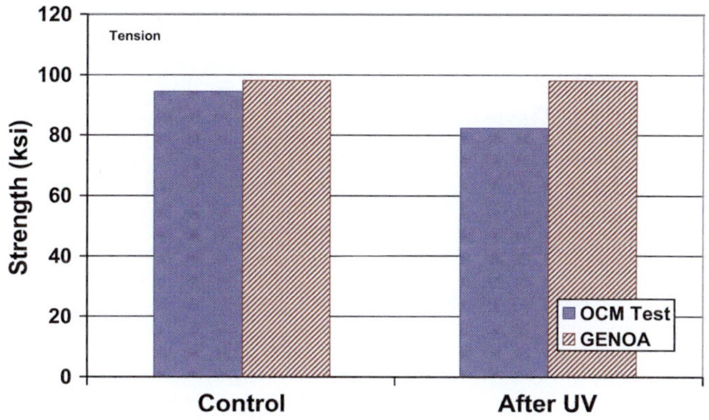

Fig. 8.23 Composite tensile strengths after 750-h UV exposure

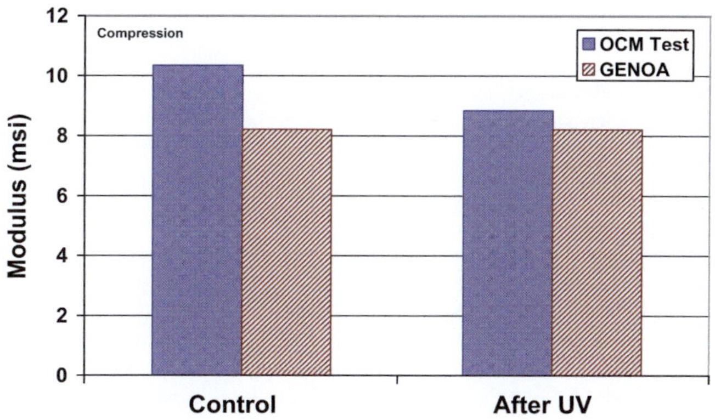

Fig. 8.24 Composite compressive moduli after 750-h UV exposure

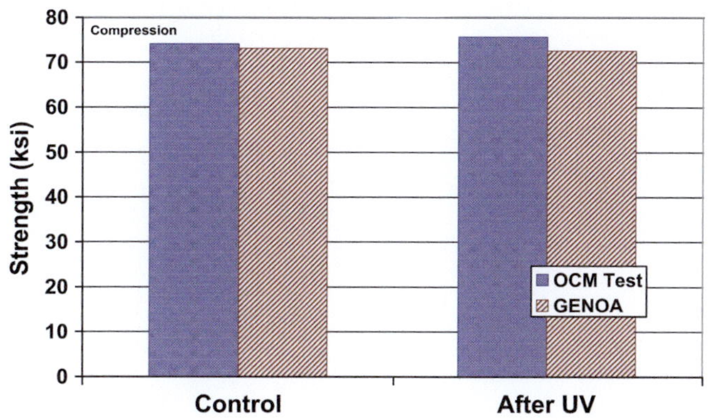

Figure 8.25 Composite compressive strengths after 750-h UV exposure

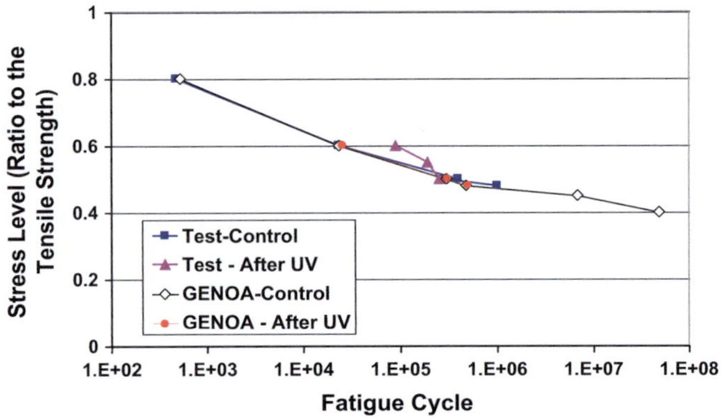

Fig. 8.26 Composite fatigue life after 750-h UV exposure

Mechanical Fatigue (R = −1) Subsequent to the UV Exposure

The tension-compression fatigue tests (R = −1) were conducted on the tri-axial composite laminate after UV exposure at three stress levels. The tested and simulated composite fatigue lives after UV exposure were compared to the corresponding control cases (Fig. 8.26). The higher fatigue life of the composite after UV exposure should be caused by the material variability or due to test scattering. GENOA simulation shows that 750-h UV exposure did not influence the composite fatigue life.

8.8 Conclusions

Effects of short term thermal cycling and ultra violet exposure on the Army mobile bridge composite materials were investigated using test and simulation methods. Results show the short-term environmental exposures do not influence the composite mechanical properties significantly. GENOA simulations, predicting composite static and fatigue responses, correlated well with test results. This demonstrates that GENOA software is a powerful tool for evaluating Army bridge material and structural performances.

Acknowledgement The authors wish to thank the US Army Research Office for the financial support through grant W911NF-05-1-0514 and Dr. Larry Russell, Jr., program manager, Environmental Sciences Division, for his continuing interest in this work.

References

1. Seemann Composites Inc., University of California at San Diego, University of Delaware, Bridge Infrastructure Renewal—Phase II. DARPA Agreement No. MDA972-94-3-0030. Vol. II, Composite Army Bridge—Task 14, March 2000
2. L. Andrady, H.S. Hamid, A. Torikai, Effects of Climate Change and UV-B on Materials. Photochem. Photobiol. Sci. **2**, 68–72 (2003)
3. F. Abdi, Z. Qian, X. Su, R. Iyer, A. Mosallam, Composite army bridge under fatigue loading. *Sampe Long Beach, CA* (2001)

Chapter 9
Modeling Creep and Relaxation Caused by Phase Dissolution

X. Li, S. Rahman, and Z.C. Grasley

Abstract Many materials are multiphasic, with an evolving, transient microstructure. If load-bearing phases within such a material dissolve while the material is under load, the stress being transmitted by those phases is handed off to neighboring phases, which leads to additional deformation. Thus, time-dependent dissolution results in creep or relaxation of the macroscopic material. To mechanistically model such dissolution-induced creep or relaxation, it is necessary to couple a model of the microstructure with the evolving states of stress and strain within the material. Here, we discuss a computationally-implemented model where creep or relaxation of an evolving composite is attributable to dissolution. Special care is taken in tracking the natural configuration of each voxel of the multiphasic material such that newly precipitated phases form in a stress-free state. The new model is utilized to model (1) relaxation of a porous material that has melting ice within its pore network and (2) relaxation due to hydration-induced dissolution within cement paste.

Keywords Microstructure • Dissolution • Computational • FEM • Viscoelastic

9.1 Introduction

Dissolution of solid load-bearing phases within multiphasic, time-evolving microstructures has been proposed as a significant mechanism leading to long-term creep or relaxation within these macroscopic materials, e.g., cementitious materials [1]. Despite its importance, to experimentally evaluate this one single mechanism towards the overall viscoelastic/viscoplastic (VE/VP) behavior of multiphasic composites is difficult. Firstly, relaxation or creep tests on long-term VE/VP behavior of composites requires long test duration and high costs; meanwhile, as multiple mechanisms likely occur simultaneously inside most composite materials (e.g., C-S-H viscoelasticity affects the VE/VP properties of cement paste over the same time scale as the dissolution effects [2, 3], it is impossible to distinguish one mechanism from the other and examine one single mechanism separately. Recently, a computational scheme that couples a microstructure evolution model and a time-stepping finite element method has been developed to simulate the overall VE/VP behavior of composites as a function of time-evolving microstructures [4, 5]. This computational scheme allows one to directly investigate the VE/VP responses of composites associated with the dissolution of solid load-bearing phases. By modifying the microstructure evolution input model within this computational scheme, the modeling approach can be generalized to predict the creep or relaxation that occurs in any composite due to microstructure evolution.

The objective of the research reported herein is to illustrate the utility of the modeling approach to predict the dissolution induced relaxation of a broad class of materials; here, we apply the model and computational scheme to both hydrating cement paste and a porous solid with ice melting within the pore network. Furthermore, the significance of the effects of phase dissolution on apparent viscoelastic relaxation is demonstrated.

X. Li • S. Rahman • Z.C. Grasley (✉)
Zachry Department of Civil Engineering, Texas A&M University, College Station, TX, USA
e-mail: xiaodanl@tamu.edu; farhanar@tamu.edu; zgrasley@civil.tamu.edu

B. Antoun (ed.), *Challenges in Mechanics of Time Dependent Materials, Volume 2*,
Conference Proceedings of the Society for Experimental Mechanics Series, DOI 10.1007/978-3-319-22443-5_9

9.2 Computationally Implemented Model

9.2.1 Conceptualization

Microstructure models for different multi-phasic materials provide input files utilized in the computational scheme. The microstructure models are capable of generating 3D snapshots for different microstructures as time evolves, and then the 3D snapshots of microstructures are meshed into voxels using a spatially aligned numerical discretization [6] such that each voxel becomes an eight node tri-linear cubic finite element consisting of a unique phase. Finite element analyses are then applied onto these 3D images to calculate the mechanical properties of these composites.

As phase change occurs from on time step to another (in the cases simulated here due either to ice melting or cement grain dissolution associated with hydration), solid phases that are transmitting stress are dissolved and replaced by other phases. A key assumption in the model is that once phase changes occur inside one microstructure, the stress originally transmitted by the dissolved solid phases are transmitted into surrounding phases in order to satisfy conservation of linear momentum. Furthermore, the newly formed phases must form in a stress-free state within the deformed configuration that conforms to the preexisting phases. This condition implies that that stress redistribution takes place much faster than chemical phase changes, and the newly formed phases carry no memories of historical responses, resulting in the production of apparent strain[1] (i.e., strain that exists independent of the state of stress) inside the microstructure. More specifically, Fig. 9.1 elaborates conceptually the process of the redistribution of stress that results stress relaxation of the macroscopic composite due to dissolution of load bearing phases. It is worth noting that such apparent VE/VP effects occur even when each microscopic phase is strictly elastic.

In the model, at the first time step, a constant periodic strain boundary condition is applied to the meshed microstructure at one particular age or time. As microstructure evolves, the microstructure for the age corresponding to a new time step would be compared to the microstructure from the previous time step. Once phase change occurs in one voxel, this voxel would be noted and assumed to be infinitely compliant at this time step such that it cannot transmit stress. Based on this assumption, finite element analysis is carried out on the overall new microstructure and the new stress and strain fields would be calculated for this time step. The model proceeds to the next time step and the voxels containing newly formed phases are assigned constitutive properties associated with the newly formed material for all subsequent steps (presuming the voxels do not change phase once again). The term "apparent VE/VP" moduli is reserved for modeling the constitutive behavior at the macroscale induced by smaller length scale mechanisms that are not due to inherent VE or VP behavior of phases within the composite material.

9.2.2 Modeling Approach

To predict the time-dependent, apparent VE/VP behavior of composites, the computational scheme was discretized in time to account for the time and (stress or strain) history dependent mechanical properties. At each time step in the finite element calculations, a 3D time-evolving microstructure generated by the microstructure model was subjected to strain-controlled periodic displacement boundary conditions, and through minimizing the total mechanical energy stored in the microstructure at each time step [4, 5],[2] the composite elastic moduli were calculated by solving the elastic equations on a regular finite element mesh [7], based on the volume averaged stress of the composite under the specific boundary condition.

For purely elastic phases subject to infinitesimal strains inside one microstructure at one time step, the total elastic energy is given by

$$En = \frac{1}{2}\int_0^1\int_0^1\int_0^1 \varepsilon_{pq} C_{pqrs} \varepsilon_{rs}\, dx\, dy\, dz, \tag{9.1}$$

---

[1] Apparent strain is defined here, from a continuum mechanics perspective, as strain that exists independent of the state of stress. Unlike other 'free' strains (e.g., strain induced by changes in temperature or moisture state of the material), the apparent strain here does not involve a change in the atomic or molecular spacing from the reference configuration. However, from a book-keeping perspective, apparent strain is treated in the same fashion as free strains.

[2] This solution procedure for the VE/VP material problem disregards any linear momentum in the body. This approach generates negligible error since the velocity of the time-dependent deformation is extremely slow under the boundary conditions considered.

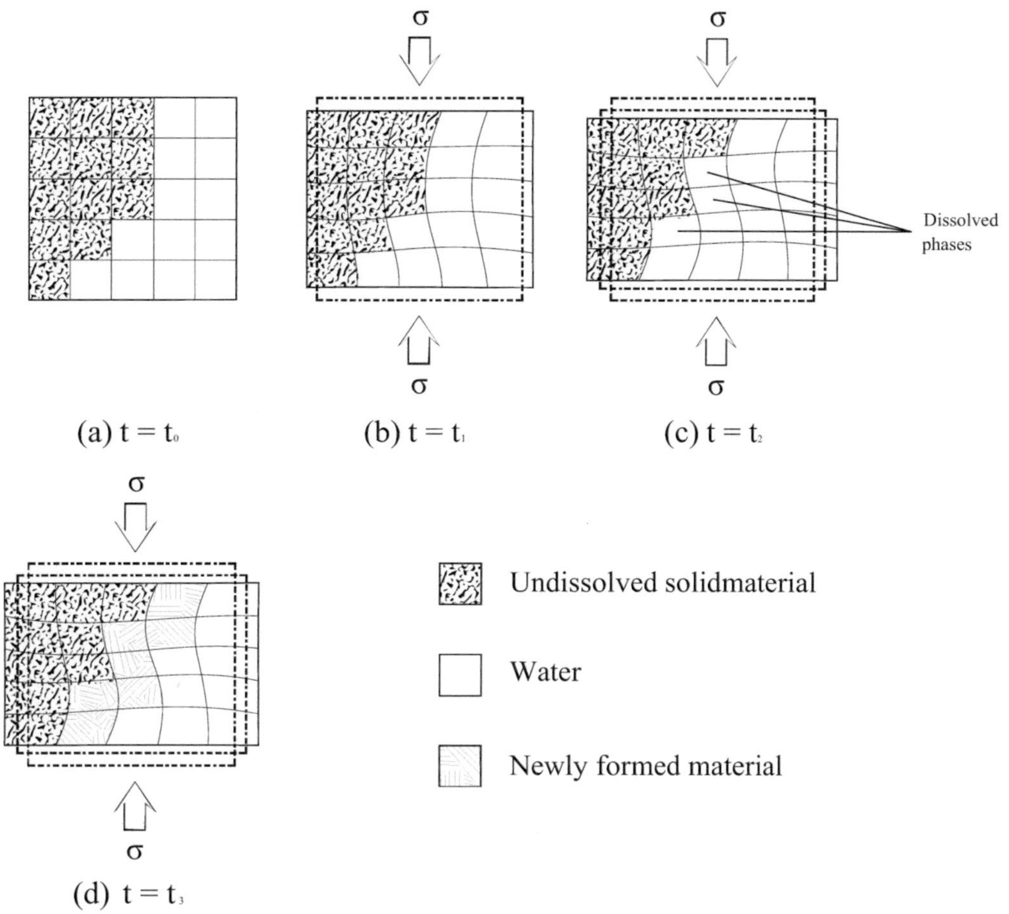

Fig. 9.1 Conceptual diagram of computational model simulating apparent viscoelastic behavior of phase-changing microstructures where σ is a constant applied external macroscale stress component and t is time elapsed. (**a**) A certain amount of load-bearing material exists before load is applied. (**b**) The body deforms elastically immediately after the application of load. (**c**) As load-bearing material dissolves, the stress on surrounding solid phases increases, which increases their elastic deformation (and thus the overall deformation of the composite microstructure). Because rate of the elastic deformation of the surrounding phases depends on the rate of phase dissolution, this results in an apparent VE/VP deformation. (**d**) New phases form in the deformed configuration in a stress-free state

where En is the elastic energy stored, ε_{pq} is the infinitesimal strain tensor and C_{pqrs} is the elastic moduli tensor; $p, q, r, s = 1$, 2, or 3, and the integral is over the volume of a single unit voxel. By expressing the strain tensor in terms of displacement components, (9.1) can be rewritten as

$$En = \frac{1}{2} u_{rp}^T D_{rpsq} u_{sq}, \tag{9.2}$$

where D_{rpsq} is the stiffness matrix and u_{rp} is the p'th component of displacement at r'th node. For any voxel inside the microstructure, the displacements of its nodes can be expressed as $u_{rp} = U_{rp} + \delta_{rp} + \delta_{rp}^{free} = U_{rp} + \delta_{rp}^{combined}$, where U_{rp} is the displacement vector determined by surrounding voxels, δ_{rp} is the correction vector determined by periodic boundary conditions and δ_{rp}^{free} is the correction vector determined by the free strain produced in the microstructure during the dissolution process. $\delta_{rp}^{combined} = \delta_{rp} + \delta_{rp}^{free}$ denotes the total displacement correction vector for a single voxel. Thus (9.2) can be rewritten as

$$En = \frac{1}{2}\left[u_{rp}^T D_{rp,sq} u_{sq} + 2\delta_{rp}^{combined} D_{rp,sq} u_{sq} + \delta_{rp}^{combined} D_{rp,sq} \delta_{sq}^{combined} \right] \tag{9.3}$$

or simplified as

$$En = \frac{1}{2} uAu + bu + C, \tag{9.4}$$

where A is the Hessian matrix comprised of the stiffness matrices, u is a vector of all the displacements, b is a constant vector, and C is a constant; both constants are determined by the external controlled strain, periodic boundary conditions, and the microstructure change during the dissolution process. When integrating the energy inside each voxel, since there is no term to be integrated that is higher order than quadratic, Simpson's rule is used for solving the exact solution of the quadratic functions.

Prescribed, periodic volumetric strain and prescribed, periodic shear strain were applied on the boundaries of a series of isotropic, evolving microstructures at different ages/times to determine the evolutions of VE/VP bulk modulus and VE/VP shear modulus. The expressions were subjected to Laplace transformations to determine the VE/VP Young's modulus. The resultant VE/VP Young's moduli were subsequently calculated via [8, 9]

$$s\bar{E} = \frac{9s\bar{K}s\bar{G}}{3s\bar{K} + s\bar{G}}, \tag{9.5}$$

where $\bar{E}, \bar{K}, \bar{G}$ represent the Laplace transforms of apparent VE/VP Young's modulus, VE/VP bulk modulus, and VE/VP shear modulus. The variable s is the Laplace transformed (time) variable.

9.3 Applications

The modeling approach was implemented to model the apparent VE/VP stress relaxation that occurs in two model materials: hydrating cement paste and a frozen porous solid that is gradually warmed. In the case of the cement paste, relaxation occurs when load-bearing cement grains dissolve in the hydration reaction. In the case of the frozen porous body, relaxation occurs due to the gradual melting of load-bearing ice crystals within the pore network. For a porous body with a distribution of fine pore sizes, thermodynamic considerations (i.e., Gibbs-Thomson eq.) cause smaller pores to be frozen and larger pores to be melted at a given temperature below 0 °C. Thus, if one has a very cold porous composite such that all pores (big and small) are frozen, places that frozen composite structure under load or displacement boundary conditions, then gradually warms the material, creep or stress relaxation will occur as portions of the pore structure gradually melt as the temperature is raised.

9.3.1 *Ice Melting in a Porous Body*

Microstructure Model

A *Mathematica* script was written to generate a three dimensional cubic microstructure with randomly distributed spherical pores of varying radii. In this algorithm, a Gaussian pore size distribution was utilized along with a selected pore volume fraction to generate an array of discrete spherical pores. The largest pores are first randomly located in a cubic box of 1 μm side length. The second largest pores are then placed randomly avoiding any overlap with the previously placed spherical pores. This process is continued until all the pores are fit into the cubic microstructure. A similar algorithm was previously applied to generate a random distribution of impenetrable objects [10] and random dispersion of cement grains [11]. The 3D image of the microstructure, as shown in Fig. 9.2, was sliced into 2D layers of 0.0028 μm and converted to grayscale. Each of these pixels in the 2D images was assigned a phase (solid skeleton/ice/water) depending on the grayscale level. Each 2D layer was later reassembled into a 3D finite element mesh with each 3D voxel representing a single phase as previously assigned.

Two microstructures were generated with two different pore size distributions: microstructure (MS1) has a narrow pore size distribution for which pore ice melts over a small range of temperature, and MS2 has a wide pore size distribution for which pore ice melts over a wide range of temperature. MS1 has a porosity (pore volume fraction) of 0.2 with a maximum and minimum pore size of 0.06 μm and 0.0062 μm, respectively. MS2 has a porosity of 0.1 and a maximum and minimum pore size of 0.0227 μm and 0.0034 μm, respectively. The frozen pore radius at which ice melts at a given temperature is determined from the Gibbs-Thomson equation according to

$$R = \frac{2\gamma_{CL}}{(T_m - T)\Sigma_m}. \tag{9.6}$$

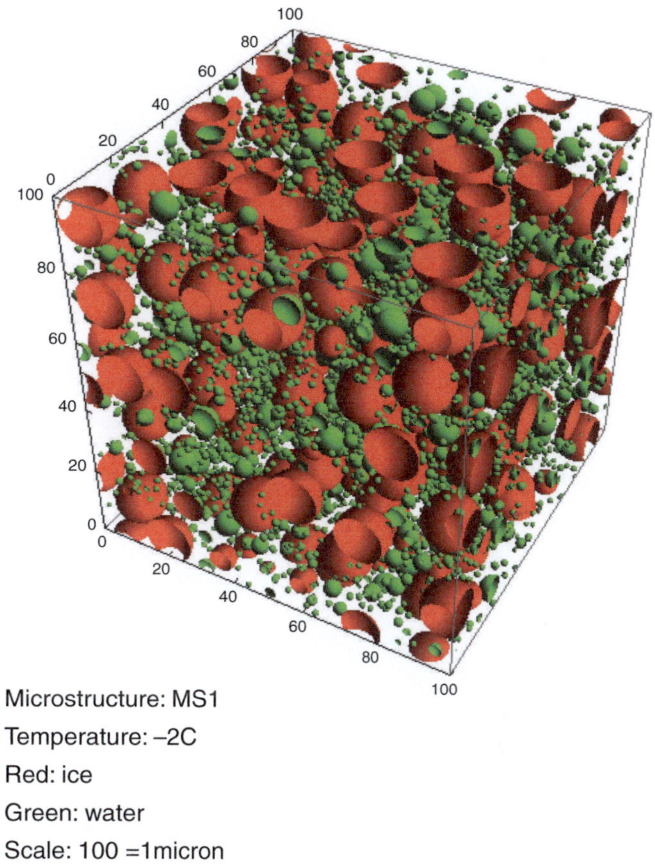

Fig. 9.2 3D microstructure of a partially frozen, porous solid. Red spheres represent frozen pores (ice) and green spheres represent melted pores with water. The solid skeleton of the porous body surrounds the pores and is transparent in this image

Here, R is the radius of the curvature of the solid–liquid interface (roughly equal to the pore size—approximated here as equal to pore size) and T is the current temperature. T_m and Σ_m are the melting point and the melting entropy, respectively. γ_{CL} is the crystal–liquid interface energy. Making use of (9.6), changes in microstructure of a frozen, water saturated porous composite with increasing temperature is presented in Fig. 9.3.

Simulation Results

Microstructures of porous materials with ice melting inside the composite were examined in the computational scheme. Different melting rates (which affects the dissolution rate of the solid load bearing phases) and different pore sizes inherent the microstructure were analyzed. For the simulations of the behavior of ice melting composites undertaken in this paper, a solid skeleton comprised of an isotropic material with elastic Young's modulus equal to 24.9 GPa and Poisson's ratio equal to 0.15 was chosen. Water phases inside the microstructure are assumed to carry zero stress, and the Young's modulus of ice is assigned to be equal to 8.7 GPa with Poisson's ratio equal to 0.31 [12]. One can expect a decreasing VE/VP Young's modulus of the composite as ice melts, and Fig. 9.4 shows how different melting rates affect the evolution of mechanical moduli of the same composite structure when the temperature increases from −10 to 0 °C. The solid line shows the evolution of the apparent VE/VP Young's modulus of the composite under the condition that temperature increases at a slower rate (1 °C/h) such that ice melts at a slower rate, while the dashed line indicates a higher rate of the warming (2 °C/h) and thus a higher melting rate. It can be seen from Fig. 9.4 that higher melting rate would induce faster relaxation of the apparent VE/VP Young's modulus due to the higher dissolution rate of the solid load-bearing ice phases.

Figure 9.5 shows how the apparent VE/VP Young's of a porous composite with different pore sizes are affected by the dissolution of ice. A material with large pore sizes establishes a larger instantaneous elastic Young's modulus, and a slower relaxation when ice starts to melt at relatively low temperatures, while a material with small pore sizes shows a faster relaxation at the beginning of the ice melting process. As temperature continues to increase there is a sudden decrease in the apparent VE/VP Young's modulus of the material with the larger pore sizes at around −2 °C at the time of 8 h, while the material with the small pore sizes continues to show a smooth relaxation curve. The major reason behind these phenomena

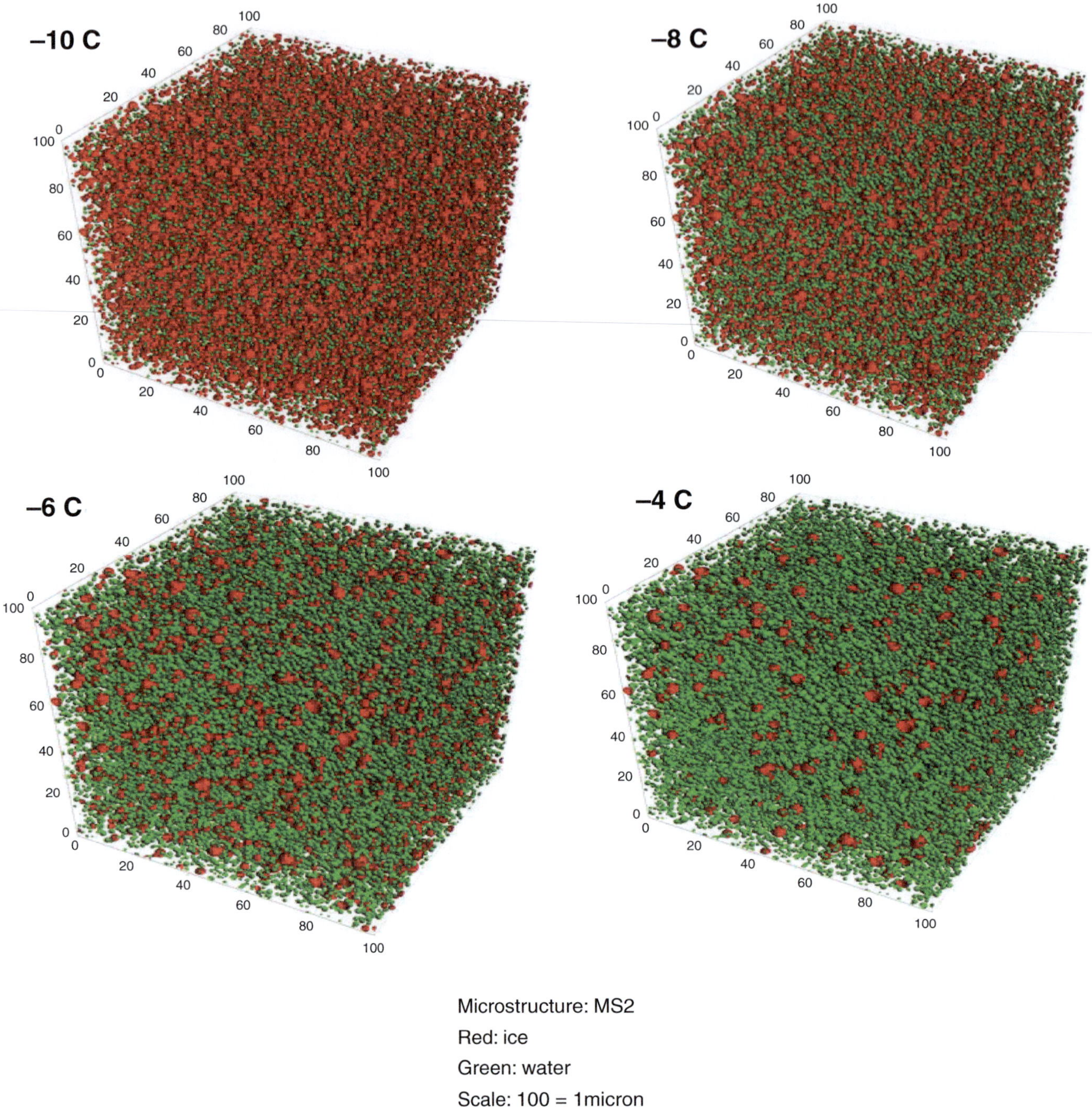

Fig. 9.3 Change in microstructure with temperature. As the temperature increases, ice in progressively larger starts to melt. The radius of these pores at a given temperature can be determined using the Gibbs-Thomson equation

is that the ice melting rate and temperature are not linearly related, but affected significantly by the pore size distribution. That is, as temperature increases, (in the simulation shown in Fig. 9.5, from −10 to 0 °C,) ice in smaller pores melts first, while the ice in larger pores only starts to melt in significant amounts after temperature increases to a certain value (in this case, around −2 °C at the 8th h for the material with larger pores). Once the threshold temperature is reached for a particular sized pore, the whole piece of ice in this pore would turn into water in a very short time, or even assumed instantaneously as it happens much faster than the temperature change. There appears to be a threshold temperature for the material with the larger pores strictly because of the narrow pore size distribution, which causes most of the pores to melt at once.

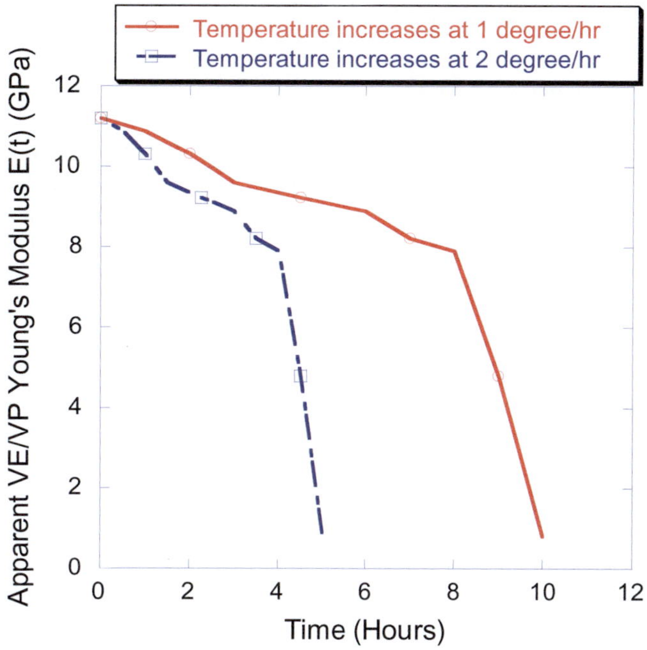

Fig. 9.4 Apparent VE/VP Young's modulus of a porous composite with ice melting inside the microstructure as it is warmed from −10 to 0 °C under different melting rates. In this graph, apparent VE/VP behavior was considered to occur strictly due to dissolution of load bearing ice phases

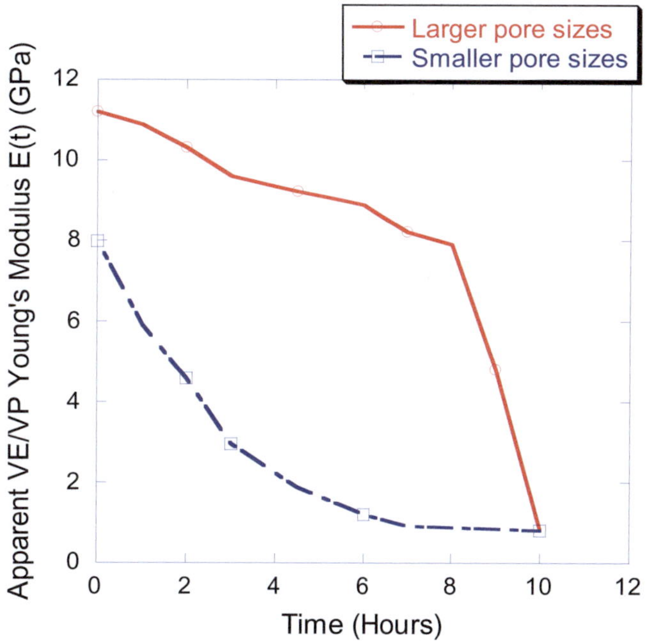

Fig. 9.5 Evolution of apparent VE/VP Young's modulus of a porous composite with different pore sizes while ice melts inside the microstructure while being warmed from −10 to 0 °C. In this graph, the temperature increases at 1 °C/h, and the apparent VE/VP behavior was considered to occur strictly due to dissolution of load bearing ice phases

It should be noted that the approximation that melted water carries no stress is simplistic. In reality, liquid water can carry significant hydrostatic stresses, and the time-dependent flow in the microstructure under induced hydrostatic pressure gradients would contribute additional relaxation associated with poromechanical effects [13–18]. Furthermore, as ice melts the density is significantly reduced, which means that the combined ice/liquid water mixture in the pore structure requires less volume, resulting in negative pressure generation in the pore fluid and potential contraction of the porous body. Such effects were neglected in this simulation since the goal was to strictly demonstrate that load redistribution effects associated with dissolution can generate substantial relaxation of a frozen porous composite.

9.3.2 Cement Paste

Microstructure Model

For cement paste, the hydration model THAMES (Thermodynamic Hydration And Microstructure Evolution) [19, 20] was used to simulate the microstructure evolution at the micrometer level during the hydration process. With the input of the original measured clinker constituent mass fractions, the particle size distribution of the cement and the water to cement mass ratio (*w*/*c*) at the sub-particle level, THAMES simulates the dissolution process of cement grains at different time steps (which represent different ages). According to empirically derived equations for clinker reaction rates [21], and based in part on phenomenological dissolution kinetics [22], THAMES calculates the pore fluid elemental composition at each time step. A thermodynamic engine GEMS (Gibbs Energy Minimization) [23, 24] was implemented in THAMES to calculate the equilibrium solution speciation and the mass of each solid constituent in equilibrium with the solution at each time step. More detailed description of the microstructure model THAMES can be found in [4, 5, 22].

Simulation Results

Microstructures of cement composites with pure elastic phases (using elastic properties taken from [22]) at different ages (from 1 to 56 days) were examined in this project. In these simulations, the relaxation of the cement pastes occurs strictly due to the time-dependent dissolution of cement grains. Figure 9.6 shows the predicted apparent VE/VP Young's modulus of 0.40 *w*/*c* cement paste under constant periodic strain boundary condition applied at different ages of 1 day and 7 days. From the predicted results, one can see that the apparent VE/VP behavior caused by dissolution of cement grains is substantial in cement paste composites. The decreasing rate of relaxation of cement paste with age (because the hydration rate slows as cement paste ages) successfully demonstrates the well-known aging effect of cement paste.

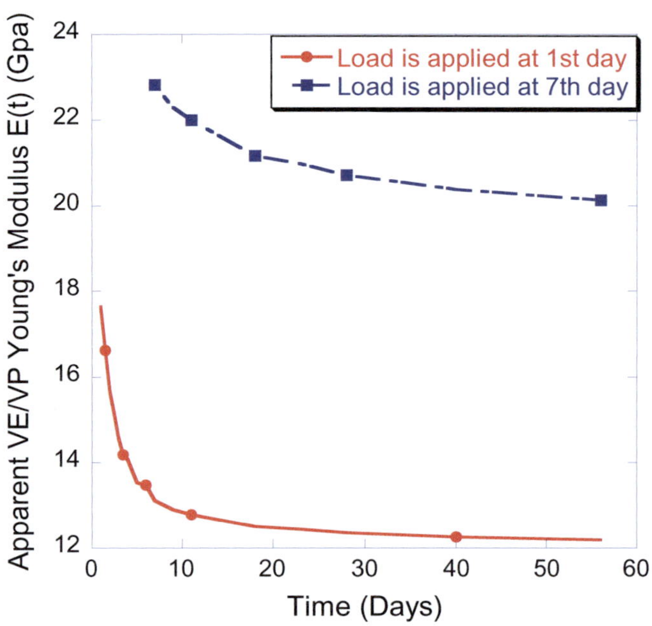

Fig. 9.6 Apparent VE/VP Young's modulus of 0.40 w/c cement paste when loaded at different ages (1 day and 7 days). In this graph, apparent VE/VP behavior was considered to occur strictly due to dissolution of load bearing cement grains. The dissolution of load bearing cement grains resulted in significant apparent VE/VP behavior for the macroscopic cement pastes, and was able to account for the well-known aging effect of VE/VP behavior of cement paste

9.4 Conclusions

A computational scheme that couples a microstructure evolution model and a time-stepping finite element method capable of tracking phase formation was developed and can be applied to different composite materials to predict their apparent VE/VP properties as a function of time-evolving microstructures. From the model simulations, the apparent VE/VP behavior of a water saturated, frozen, porous composite due to dissolution of solid ice phases upon warming is significant. The model captures the effect of heating rate and pore size distribution on the simulated relaxation rate. For cement paste, simulation results indicate that dissolution of cement grains of hydrating cement paste is a significant factor in the overall early-age creep and relaxation of the paste and is capable of demonstrating the well-known aging effect of cement paste.

Acknowledgments This research was supported by the National Science Foundation CAREER Award Program under grant number CMMI-0843979. Any opinions, findings, and conclusions or recommendations expressed in this material are those of the author(s) and do not necessarily reflect the views of the National Science Foundation.

References

1. M. Suter, G. Benipal, Constitutive model for aging thermoviscoelasticity of reacting concrete I: theoretical formulation. Mech. Time-Depend. Mater. **14**, 277–290 (2010)
2. O.M. Jensen, P.F. Hansen, Autogenous deformation and change of the relative humidity in silica fume-modified cement paste. ACI Mater. J. **93**, 539–543 (1996)
3. B.T. Tamtsia, J.J. Beaudoin, Basic creep of hardened cement paste. A re-examination of the role of water. Cem. Concr. Res. **30**, 1465–1475 (2000)
4. X. Li, Z.C. Grasley, E.J. Garboczi, J.W. Bullard, Modeling the apparent and intrinsic viscoelastic relaxation of hydrating cement paste. Cem. Concr. Compos. **55**, 322–330 (2015)
5. X. Li, Z.C. Grasley, E.J. Garboczi, J.W. Bullard, Computing the time evolution of the apparent viscoelastic/viscoplastic Poisson's ratio of hydrating cement paste. Cem. Concr. Compos. **56**, 121–133 (2015)
6. W. Ruetz, A hypothesis for the creep of hardened cement paste and the influence of simultaneous shrinkage, Proceedings of International Conference on the Structure of Concrete, (1968)
7. T.I. Zohdi, *Homogenization Methods and Multiscale Modeling* (Wiley, Chichester, 2004)
8. R.B. Bohn, E.J. Garboczi, User Manual for Finite Element and Finite Difference Programs: A Parallel Version of NISTIR-6269, U.S. Department of commerce, Technology Administration, National Institute of Standards and Technology, Information Technology Laboratory, Building and Fire Research Laboratory, (2003)
9. C.J. Haecker, E.J. Garboczi, J.W. Bullard, R.B. Bohn, Z. Sun, S.P. Shah, T. Voigt, Modeling the linear elastic properties of Portland cement paste. Cem. Concr. Res. **35**, 1948–1960 (2005)
10. E.J. Garboczi, A.R. Day, An algorithm for computing the effective linear elastic properties of heterogeneous materials: three-dimensional results for composites with equal phase Poisson ratios. J. Mech. Phys. Solids **43**, 1349–1362 (1995)
11. S.P. Timnoshenko, J.N. Goodier, *Theory of Elasticity* (McGraw-Hill, New York, 1970)
12. P.V. Hobbs, *Ice Physics* (Oxford University Press, Oxford, 2010)
13. S. Rahman, Z. Grasley, A poromechanical model of freezing concrete to elucidate damage mechanisms associated with substandard aggregates. Cem. Concr. Res. **55**, 88–101 (2014)
14. C.K. Leung, Z.C. Grasley, Poromechanical damping of cementitious materials. J. Mater. Civil Eng. **24**, 232–238 (2012)
15. C.A. Jones, Z.C. Grasley, Correlation of radial flow-through and hollow cylinder dynamic pressurization test for measuring permeability. J. Mater. Civil Eng. **21**, 594–600 (2009)
16. C.A. Jones, Z.C. Grasley, Correlation of hollow and solid cylinder dynamic pressurization tests for measuring permeability. Cem. Concr. Res. **39**, 345–352 (2009)
17. C.A. Jones, Z.C. Grasley, Novel and flexible dual permeability measurement device for cementitious materials. ACI Mater. J. **106**, 192–197 (2009)
18. C.A. Jones, Z.C. Grasley, Measuring Concrete Permeability Using Dynamic Pressurization: Achieving Saturation, 2009 NRMCA Concrete Technology Forum, (electronic proceedings) T05.2, (2009)
19. L.D. Landau, E.M. Lifshitz, *Theory of elasticity*, 3rd edn. (Pergamon Press, Oxford, 1986)
20. D. Bentz, E. Garboczi, K. Snyder, Hard core/soft shell microstructural model for studying percolation and transport in three-dimensional composite media. NISTIR **55**, 62–65 (1999)
21. A. Yazdanbakhsh, Production, characterization, and mechanical behavior of cementitious materials incorporating carbon nanofibers. (Texas A&M University, 2012)
22. J.W. Bullard, B. Lothenbach, P.E. Stutzman, K.A. Snyder, Coupling thermodynamics and digital image models to simulate hydration and microstructure development of Portland cement pastes. J. Mater. Res. **26**, 609–622 (2011)
23. B. Lothenbach, F. Winnefeld, Thermodynamic modelling of the hydration of Portland cement. Cem. Concr. Res. **36**, 209–226 (2006)
24. L. Parrot, D. Killoh, Prediction of cement hydration. Proc. Br. Ceram. Soc. **35**, 41–53 (1984)

Chapter 10
Effects of Net and Solid Skins on Self-Supporting Lattice Structures

A.O. Aremu, Ian Maskery, C.J. Tuck, I.A. Ashcroft, R.D. Wildman, and R.J.M. Hague

Abstract Multi-functional capabilities of Lattice materials allow them to be used in weight bearing applications, impact absorption and heat dissipation. Previously, the range of cellular materials was limited by constraints in traditional manufacturing technologies. This situation has been mitigated by advances in additive manufacturing (AM) techniques, which allows the manufacture of complex parts directly from three dimensional CAD models. However, other manufacturing difficulties emerge with AM techniques. Such difficulties exist with metallic components produced with selective laser melting (SLM). The need for support could compromise the quality of the lattice being built since it is difficult to remove such structures. Lattices requiring little or no support are better suited for SLM. A method was recently developed to fit such lattices to complicated three dimensional domains and consist of an approach to improve the performance of the trimmed lattices with solid and net skins. In this paper, we investigate the influence of these skins on the stiffness of Body-Centred Cubic (BCC) and Double-Gyroid (Dgyroid) lattices via finite element analysis. Including a net skin on the BCC lattice improved its stiffness as the thickness of the skin was increased. The stiffness of Dgyroid lattice was insensitive to a net skin. However, solid skins improve the performance of both lattice types.

Keywords Lattice • Net-skin • Finite element • Additive manufacturing • Stiffness

10.1 Introduction

The realization of novel designs can be limited by the manufacturing route. This is due to the inherent constraints within traditional manufacturing techniques which necessitate the subjection of component designs to rules that minimize manufacturing and assembling difficulties [1, 2]. Dominance of traditional manufacturing techniques such as injection moulding, machining or stamping is based on their ability to achieve mass production at a relatively low unit cost. For smaller production quantities where product customization is important, these techniques become significantly expensive due to the high cost of tooling. A potential route to overcoming this is Additive Manufacturing (AM), which is a flexible approach to manufacturing. AM techniques, such as selective laser melting (SLM) are increasingly being used to realize high value metallic components, especially where lattice structures are embedded in the components to improve specific performance of the components. It is important to note that SLM does have design constraints, with the need for supports being the most paramount. Not all geometries require supports since those with features that are inclined at angles around 45° and above to the build platform will self-support. Removal of supports from lattices could be difficult, deteriorate surface finish, and time consuming. It is therefore preferable to consider lattices that don't require support when designing for SLM.

Lattice structures can be used to bear weight, absorb impact or dissipate heat. A broad range of lattice topologies have been studied in the past, with only a few associated with SLM. Previous works on lattices manufactured with SLM focused on self-supporting lattices like Body-Centred Cubic (BCC) or Gyroid lattices and their variants as seen in [3–7]. Investigating the performance of other of self-supporting lattice is beneficial. One of such is a variant of the Gyroid lattice described in [7] as the Double Gyroid (Dgyroid) lattice. The structural features of the Dgyroid lattice make it potentially useful for SLM parts. To investigate this, suitable lattice generation techniques are required to incorporate this lattice in the parts. Brennan-

A.O. Aremu (✉) • I. Maskery • C.J. Tuck • I.A. Ashcroft • R.D. Wildman • R.J.M. Hague
Faculty of Engineering, EPSRC Centre for Innovative Manufacturing in Additive Manufacturing,
University of Nottingham, Nottingham NG7 2RD, UK
e-mail: adedeji.aremu@nottingham.ac.uk

The original version of the chapter has been revised: A correction to this chapter can be found at https://doi.org/ 10.1007/978-3-319-22443-5_13

B. Antoun (ed.), *Challenges in Mechanics of Time Dependent Materials, Volume 2*,
Conference Proceedings of the Society for Experimental Mechanics Series, DOI 10.1007/978-3-319-22443-5_10

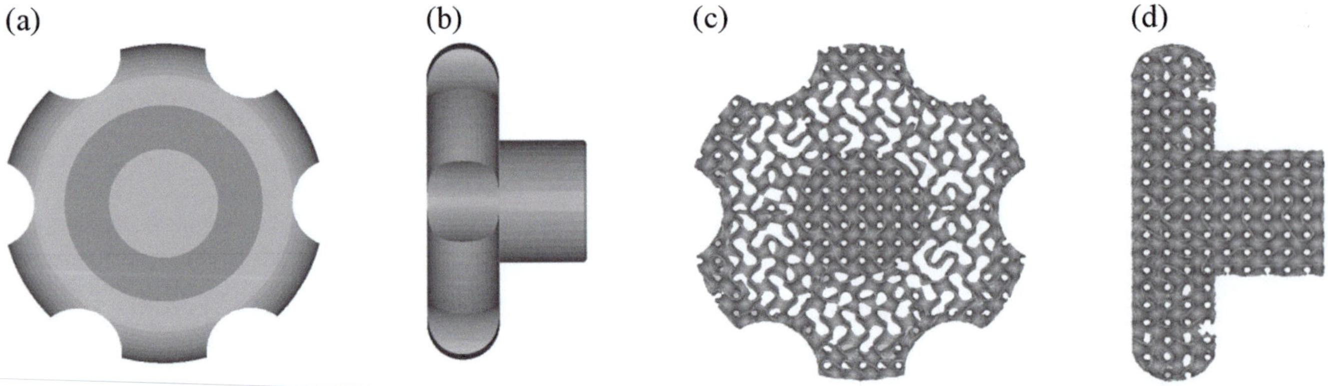

Fig. 10.1 An illustration of a trimming based approach to lattice generation (**a**) Front view, screw handle (**b**) Side view, screw handle (**c**) Front view, trimmed lattice (**d**) Side view, trimmed lattice

Fig. 10.2 A sectional view of a trimmed lattice with a solid skin

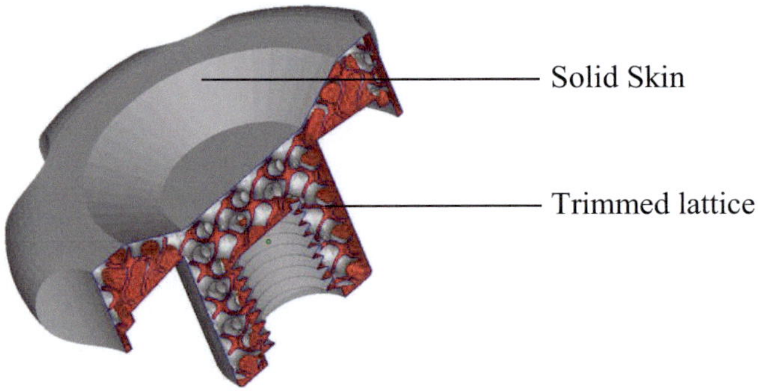

Craddock [8] describes methods that are used to fit lattices to complicated domains, identifying also the versatility of using a trimming based approach to construct lattices. Figure 10.1 illustrates the trimming approach to lattice construction. This approach involves the tessellation of a unit cell over the lattice domain. Regions of the lattice residing in the interior of the domain are retained while others are removed.

The screw-handle shown in Fig. 10.1a, b represents the domain used to trim a tessellation of the Dgyroid lattice [9] to derive the structure shown in Fig. 10.1c, d. Before tessellation, it is preferable to convert the unit cell and the domain to voxel models to enhance the flexibility of the lattice generation. An artefact of a trimming approach is the existence of unconnected features at the boundary of the trimmed lattice. This weakens the lattice and could initiate failure in service. Including a solid skin could avert the problem, however, this could,

- hinder the removal of unprocessed powder particles after production,
- limit access to the high surface region of the lattice (if the lattice is used for thermal dissipation),
- Increase the mass of the entire structure.

The trimmed lattice in Fig. 10.1c, d is without skin, Fig. 10.2 shows a sectional view of this same lattice with a solid skin. A porous or net-skin is more beneficial since it serves to connect the hanging features of lattice while minimizing the skin's mass. The topology of the skin should reflect that of the trimmed lattice to ensure the hanging features are connected.

This can be achieved by projecting features in the vicinity of the trimmed lattice boundary into the domain occupied by a solid skin. Voxel models offer great flexibility in implementing this strategy, this paper uses such models to compare the effects of solid and net skin on the stiffness of the BCC and Dgyroid Lattices. Two cell sizes and different skin thickness are considered in the. STL-files were constructed for the cells and then voxelized. Both cells were fixed at the same density as explained in the next section. Voxel models of the cells were used to construct the trimmed lattice and analysis models.

10.2 Body Centred Cubic and Double Gyroid Lattices

The BCC unit cell is composed of eight cylindrical members that connect the corners of a cube to the centre Fig. 10.3a. For a uniformly sized lattice, the radii, *R*, of these members are equal and its values determine the density of the cell. For functionally graded lattices, *R* may vary across the design domain. We focus on the lattices with uniform cellular features. The struts are inclined at 45° with respect to the build direction, or z-axis. Therefore, the BCC lattice is self-supporting. The ends of the cylindrical member are trimmed to restrict them to the boundaries of the cube.

The Dgyroid (Fig. 10.2a and Dgyroid) unit cell is a variant of the Gyroid unit cell [9]; which belongs to a family of minimal surfaces [3]. Dgyroid conforms to the equation:

$$\left(\sin\left(\frac{2}{L}x\right)\cos\left(\frac{2}{L}y\right)+\sin\left(\frac{2}{L}y\right)\cos\left(\frac{2}{L}z\right)+\sin\left(\frac{2}{L}z\right)\cos\left(\frac{2}{L}x\right)\right)^2=t^2 \quad (10.1)$$

x, y, and z, are Cartesian coordinates in a three dimensional space. L is the length of the cube while $0 \le t \le 1.413$ and controls the density of the Dgyroid cell. The curvature of the Dgyroid cell allows it to self-support on the SLM. Work in this paper is focused on the performance of the Dgyroid lattice in its matrix phase (Fig. 10.3b) rather than the network phase shown in Fig. 10.3c. Both of which conform to (10.1) as seen in [7]. To confirm both the BCC and Dgyroid lattices self-support on SLM, lattices of varying unit cell types were constructed and manufactured with Renishaw's AM250 machine. Images of these lattices are shown in Fig. 10.4.

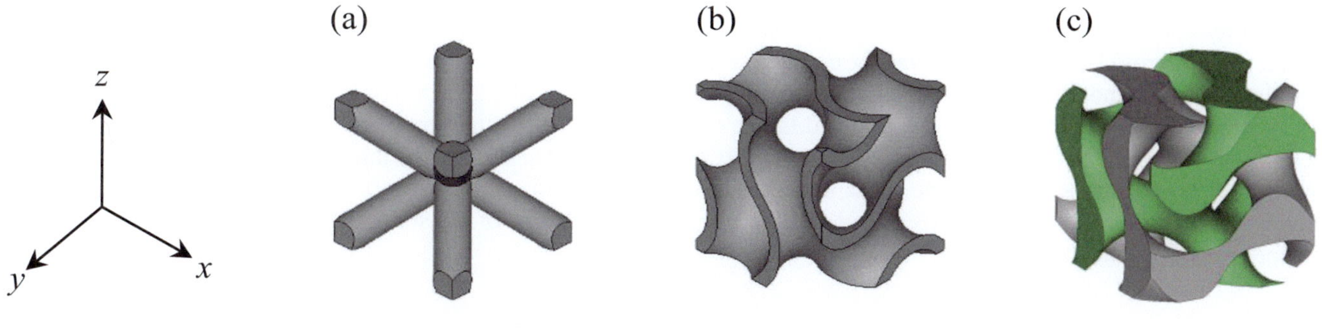

Fig. 10.3 Self-supporting unit cells (**a**) BCC (**b**) Matrix Phase of Dgyroid (**c**) Network phase of Dgyroid

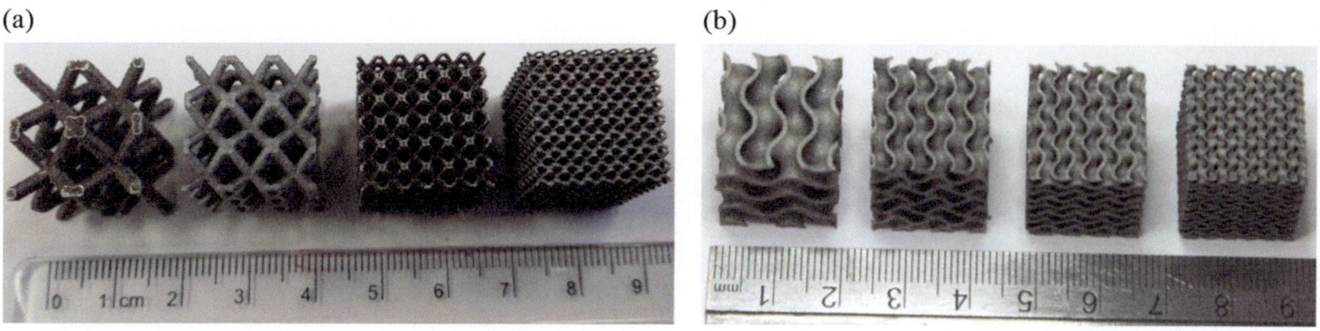

Fig. 10.4 Manufactured self-supporting lattices with different unit cell sizes, 9 mm, 6 mm, 4.5 mm and 3 mm (**a**) BCC (**b**) Double Gyroid

10.3 Lattice and Net-Skin Generation

A voxel based approach to lattice and net skin generation proposed in [10] was used to construct the lattices and their skins. The first step was to create STL files for both BCC and Dgyroid cells with a MATLAB script to allow better control of the cellular densities; rather than via CAD software since it was difficult to predetermine values for cellular parameters that would give the target density. The density represents the ratio of the volume of the lattice to that of the domain it occupies and was set at 0.24. R of the cylindrical members in the BCC cell was assumed to be equal. An initial study was performed to determine the polynomial equations that relate BCC cellular parameter, R and Dgyroid parameter t to their corresponding densities, d. R was related to d according to:

$$R = -305.68d^6 + 482.23d^5 - 323d^4 + 121.69d^3 - 29.27.68d^2 + 7.19d + 0.21 \tag{10.2}$$

While t relates to d according to:

$$t = 13.39d^6 - 26.83d^5 + 22.40d^4 - 10.16d^3 + 2.63d^2 + 1.16d + 0.03 \tag{10.3}$$

These equations were then used to determine the R and t that corresponds to 0.24 densities for both the cells. R was determined to be 1.18 mm while t was 0.3751. An array holding the Cartesian coordinates of a point cloud was initiated within a three dimension space. Points existing on the surface of the cylindrical members of the BCC cell were at a distance R from the central axis of these members while those obeying (10.1) defined the topology of the Dgyroid cell. Both sets of points were extracted and used to construct STL facets.

STL-file for both cells and a domain were subsequently voxelized, a process that constructs a discrete three dimensional array from the triangular facets [8]. This involves initiating an empty three dimensional array, setting regions residing on the interior and boundary of the STL geometry to 1 while those on the exterior are set to zero. Detailed description of classifying the voxels this way can be seen in [10, 11]. Voxel models of the cells were tessellated over the domain of the lattice and then trimmed to fit the domain as explained in [8, 10]. The tesselated unit cell were fitted to a cube with dimensions shown in Fig. 10.5a.

Net skins were constructed by projecting an image of the trimmed lattice onto the boundary of the lattice domain. This ensured that unconnected features were linked by the skin. Two different methods were used to construct the skins owing to contrasting features of the BCC and Dgyroid lattices. Net skin for the BCC Lattice were constructed by projecting voxels in the vicinity of the boundary along the three Cartesian axes as described in [10]. A second approach was used to construct the skin for the Dgyroid lattice. This involved the extraction of a net-plane within the topology of the Dgyroid lattice. The closest net-plane to the boundary (17.45 mm into the lattice) of the Dgyroid lattice along the x-axis is shown in Fig. 10.5b. This plane was projected unto the boundary of the trimmed lattice and combined with other net-plane in the y and z axes to form the net. Figure 10.6 shows front and isometric views of the BCC and Dgyroid net skins. The thickness of the skins was controlled by further projecting the skin into the domain of the structure. Solid skins were constructed by extracting the outermost voxels in the domain using MATLAB's '*bwperim*' command. Voxel arrays of the solid and net skins were combined and mapped into a hexahedral mesh.

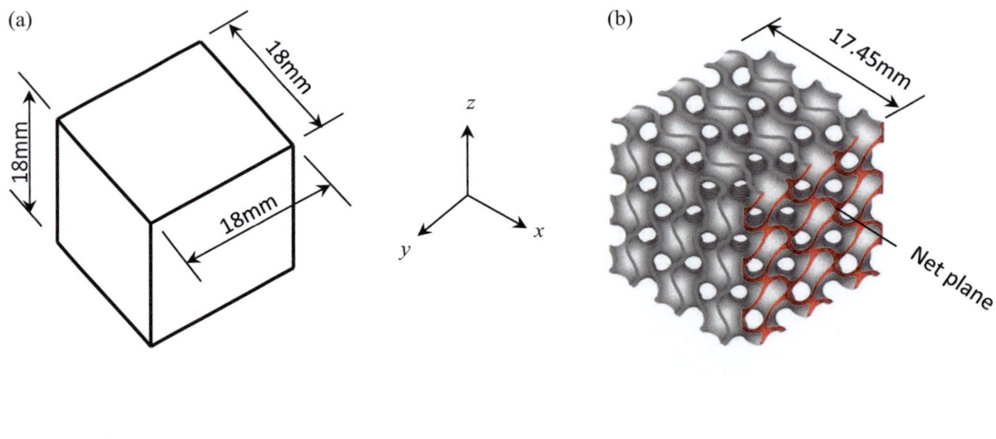

Fig. 10.5 Lattice domain and sectional net plane (**a**) Lattice domain (**b**) Net plane for Dgyroid lattice

Fig. 10.6 Net-Skins (**a**) BCC, front view (**b**) BCC, isometric view (**c**) Dgyroid, front view (**d**) Dgyroid, isometric view

10.4 Finite Element Study

A finite element analysis (FEA) was performed to determine the effects of the net and solid skins on the structural stiffness of the BCC and Dgyroid lattice. Young's modulus and Poisson's ratio of elements of the lattice was assumed to be 62 GPa and 0.33 GPa respectively. A 1 mm thick plate was placed on all lattices and loaded with compressive pressure load of 0.93 MPa (Fig. 10.7) corresponding to a total force of 300 N on the top surface of the plate. Material properties of the plate were assumed to be many times greater than that of the lattice. Unit cells size was set at 6 and 9 mm and convergence study was initially performed to determine the appropriate mesh size that minimized FEA errors. This was determined to be approximately 900,000 hexahedral elements. The skin thickness was set at 0, 0.125, 0.25, 0.50 and 0.625 mm for both net and solid-skins. Results are presented and discussed in the next section.

10.5 Results and Discussion

Figure 10.8 shows a plot of displacement against skin thickness. Data intercepting the vertical axis represent those without the skins. For both net and solid skins, the Dgyroid lattices performed better than the BCC. Also the size of the cells had minimal effects on performance as shown in Fig. 10.8a, b. While the Dgyroid lattice is insensitive to the thickness of the net skin as shown in Fig. 10.8a, the BCC lattice becomes stiffer as the skin thickness is increased.

These structural behaviours are a combined effect of different skin topologies and the intrinsic behaviour of the lattices. Since the Dgyroid lattice without a skin performed better than the BCC lattice with 0.625 mm net-skin, it can be inferred that Dgyroid offers higher specific stiffness. The solid skin influenced both lattice types in a similar manner, though it can be seen that this skin had a greater effect on the BCC lattice, reducing the displacement from 0.010 mm to less than 0.002 mm as the thickness is increase from 0 to 0.0625 mm. The displacement range for the Dgyroid lattice is less broad with a range of 0.0035 to less than 0.0015 for same skin thickness. These suggest that the Dgyroid lattice is relatively insensitive to the inclusion of the solid skin. Much of this insensitivity to skins can therefore be due to intrinsic properties of the Dgyroid lattice since the solid skin has the same topology for both BCC and Dgyroid lattice.

As mentioned earlier, the densities of the lattices without the skins were fixed at 0.24. Including the skins increased the densities and caused a variation in the mass of the lattices. This is a consequence of the two different methods used to generate the net skins and the contrasting interaction between the boundaries of the unskinned lattice and those of the solid skins. The stiffness to volume ratio, *K* was calculated and used to compare the performance of the lattices according to:

$$k = \frac{-f}{uV} \tag{10.4}$$

where *f* is the total force on the plate, *u* is the *z*-displacement and *V* is the volume of the lattice. A plot of *K* against skin thickness in shown in Fig. 10.9. Observations in the displacement plot shown in Fig. 10.8 appear to be consistent with those in Fig. 10.9. This implies that the contrast in volume did not contribute to the behaviour described previously. The effects of cell size are minimal for both net and solid skins. Also, Dgyroid has superior *K* to those of BCC (Fig. 10.9a) and show lower

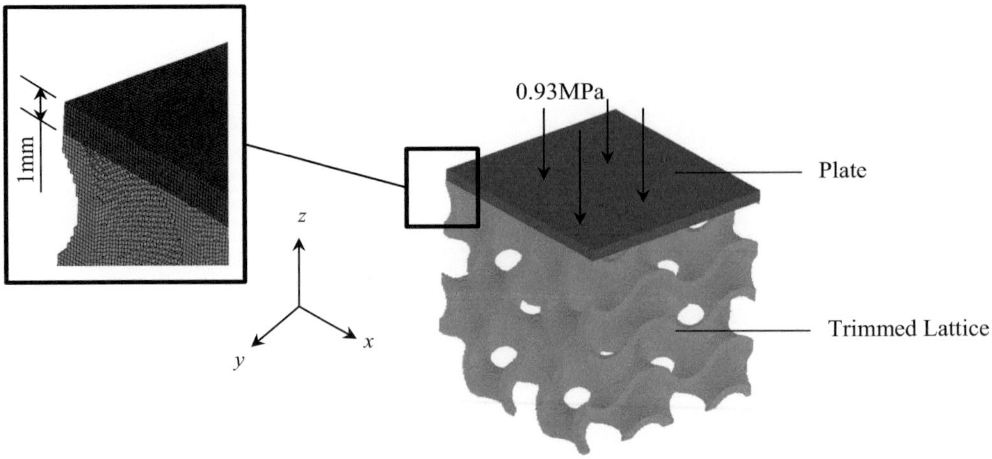

Fig. 10.7 Finite element model using hexahedral elements and the application of compressive pressure load on the lattices

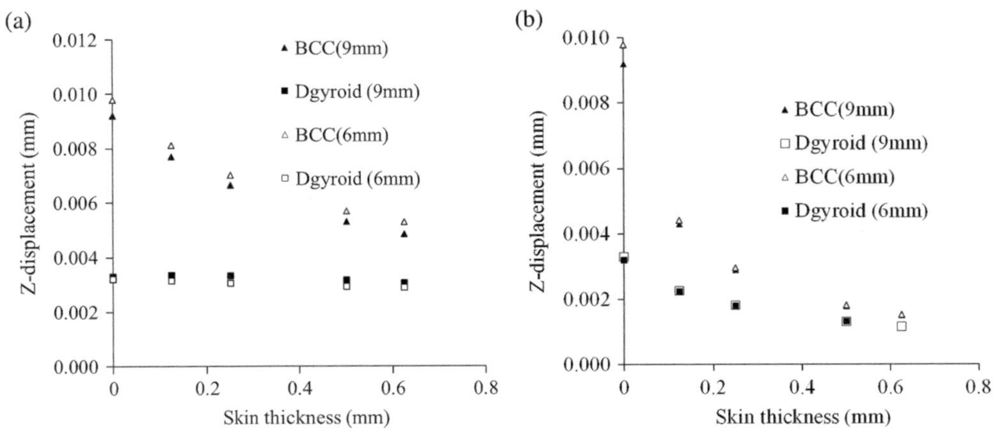

Fig. 10.8 Displacements against skin thickness (**a**) Net Skin (**b**) Solid Skin

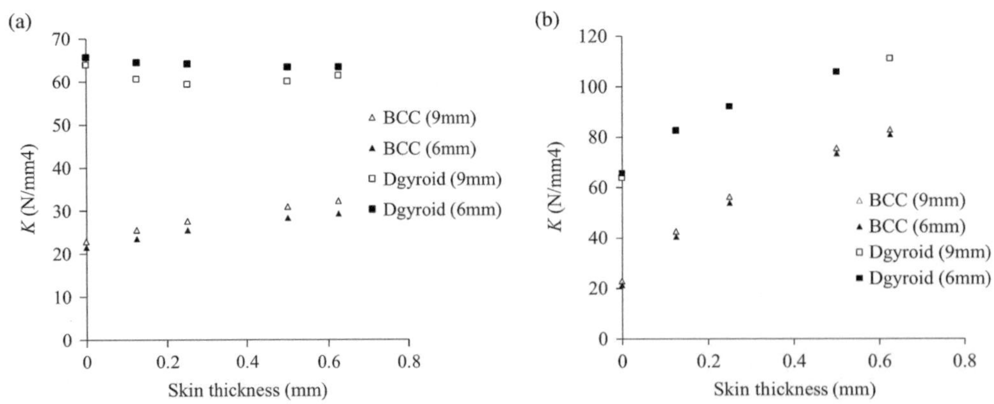

Fig. 10.9 *K* against skin thickness (**a**) Net Skin (**b**) Solid Skin

K sensitivity to the net skin unlike the BCC which is stiffer as the skin thickness is increased. The solid skin had the same effect on K for both lattice types which increased in a similar manner as seen in Fig. 10.9b. The BCC lattice with a solid skin performs better than the Dgyroid without a skin only after the skin thickness reached 0.25 mm.

Generally, lattices with the solid skins have higher K than those of the net skins. However, the Dgyroid lattice without the skin is able to achieve 57 % of the maximum K in Fig. 10.9b, BCC which is only able to achieve 27 % of this K value. Practical parts, hosting these lattices are more likely to have both skin types; with the walls of the part in contact with the lattice being solid while others boundaries of the lattice are net. The walls of the part would improve the performance of the lattice while the net skin could either contribute to this or be passive (depending on cell type), and offer access to the lattice region either during post processing or service. Applications involving the Dgyroid lattice should exclude the net skin to avoid excess weight in the part.

10.6 Summary and Conclusions

A finite element study was performed to compare the behaviour of the BCC and the Dgyroid lattice with both net and solid skins. The ability to manufacture these lattices with the selective laser melting technique makes an investigation into their structural properties useful. Lattice structures and their skins are generated via a voxel based approach which was subsequently used to define the analysis model. The Dgyroid lattice was seen to be relatively insensitive to the thickness of the net skin for the range of skin thickness considered. However, its stiffness to volume ratio increased with an increase in solid skin thickness. The performance of the BCC lattice improved with the existence of a skin and became better as the skin thickness was increased. However, heavier BCC lattice with the net skin were inferior to The Dgyroid lattice without a skin. Also, there was minimal effect of cell size on the performance of the lattices. These results suggest the Dgyroid is superior to the BCC lattice.

Acknowledgement This work is funded by the engineering and physical science research council (EPSRC) and the Innovate UK.

References

1. G. Boothroyd, P. Dewhurst, W.A. Knight, *Product Design for Manufacture and Assembly* (CRC, New York, 2011)
2. C. Chu, G. Graf, D.W. Rosen, Design for additive manufacture of cellular structure, Comput. Aided Des. Applic. p. 11 (2008)
3. R. Hasan, R. Mines, P. Fox, Characterization of selective laser melted Ti-6Al-4 micro lattice struts. Proc. Eng. **10**, 536–541 (2011)
4. R. Gumruk, R.A.W. Mines, Compressive behaviour of stainless steel micro-lattice structures. Int. J. Mech. Sci. **68**, 125–139 (2013)
5. C. Yan, L. Hao, A. Hussein, P. Young, D. Raymont, Evaluations of cellular lattice structures manufactured using selective laser melting. Int. J. Mach. Tool Manuf. **65**, 32–38 (2012)
6. M. Smith, Z. Guan, W.J. Cantwell, Finite element modelling of the compressive response of lattice structures manufactured using the selective laser melting technique. International Journal of Mechanical Science **67**, 28–41 (2013)
7. M.R.J. Scherer, Double-gyroid-structured functional materials. Springer Thesis. (Springer, Switzerland, 2013). 10.1007/978-3-319-00354-2_2
8. J.P.J. Brennan Craddock, The investigation of a method to generate conformal lattice structures for additive manufacturing, PhD. thesis, Loughborough University (2011)
9. A.H. Schoen, Infinite periodic minimal surfaces without self-intersections, NASA Technical Note D-5541 (1970)
10. A. Karabassi, G. Papaioannou, T. Theoharis, A depth buffer based voxelization algorithm. J. Graphics Tools **4**(4), 5–10 (1999)
11. S. Fang, H. Chen, *Hardware Accelerated Voxelization*. Volume Graphics (Springer, New York, 2000), pp. 301–315

Chapter 11
Hybrid Joining Through Additive Manufacturing

Thao Gibson, G.P. Tandon, Adam Hicks, John Middendorf, Bryce Laycock, and Gerard Simon

Abstract Current methods for compactly joining High Temperature Polymer Matrix Composites (HTPMC's) and metals used in engine components often rely on structural adhesives. These methods have inherent challenges such as verification of the bond quality, strength and temperature limitation of the polymer in the adhesive, interfacial adhesion strength and wetting between the metal and polymer, and a lack of load transfer via fiber reinforcement across the joint or via interlocking.

This work will discuss efforts to develop an innovative solution for hybrid joining of a metal to HTPMC using state-of-the art additive manufacturing and processing techniques. Ti-6Al-4V powder is deposited using the Selective Laser Melting (SLM) process onto carbon fiber fabrics, and the fabrics' bottom layer is then infused with a polymer. Ultimately, the hybrid composite consists of a fully densified metal on one face and an HTPMC on the opposing face, with a tailored transition region containing a significant amount of fiber reinforcement connecting the two faces. This hybrid material will be tested to establish its mechanical and physical properties on a small scale, and its microstructure will be characterized as the overall scope of this work. This paper is focused on the first part of the hybrid whereas the deposition of Ti-6Al-4V powder onto carbon fabrics was achieved with optimum processing parameter using SLM method.

Keywords Hybrid • Polymer matrix composites • Additive manufacturing • Selective Laser Melting

11.1 Introduction

Traditional composite sandwich structure hybrids, in which the polymeric composite is sandwiched between two layers of material composed of metal or ceramic, can yield hybrids that yield metal-composite-metal, ceramic-composite-ceramic, or metal-composite-ceramic structures. These hybrid structures have advantages over traditional polymeric composites. For example, a metal-polymeric composite hybrid laminate where the composite is sandwiched between two titanium (Ti) thin sheets showed increased temperature uses, improved thermal oxidation stability, and better damage tolerance of aircraft skin [1]. Yet, the usage of an adhesive between a metal and polymer matrix composite (PMC) leads to thermal expansion and failure, inducing stress due to the coefficient of thermal expansion (CTE) missmatch, which is a primary concern of this hybrid structure. Therefore, the solutions focusing on addressing hybrid joining of a metal to PMC using state-of-the art additive manufacturing and composite processing techniques, thus eliminating an adhesive, are the major focus of our research in order to successfully implement hybrid materials into aircraft. Our approach for a hybrid structure containing a dense metal layer, metal infiltrated onto carbon fabric (metal + fiber + resin), and polymeric composite layer is shown in Fig. 11.1.

The purpose of this work is to build a hybrid prototype that can also replace the traditional adhesives in between the metal and polymeric composite, which is one of the weak points in metal and polymer composites today. Using SLS process, our goal is to yield a Ti metal and carbon fiber hybrid which is fully dense of metal on one side and has dry fiber on other side. The polymer can infuse into the dry fiber side of the structure and metals can be directly welded to the other side. This should

T. Gibson • G.P. Tandon (✉)
University of Dayton Research Institute (UDRI), 300 College Park Avenue, Dayton, OH 45469-0060, USA

Air Force Research Laboratory (AFRL/RXCC), 2941 Hobson Way, Wright Patterson Air Force Base, OH 45433-7750, USA
e-mail: Gyaneshwar.tandon.ctr@us.af.mil

A. Hicks • J. Middendorf • B. Laycock
Mound Laser Photonic Center (MLPC), 2941 College Drive, Kettering, OH 45420, USA

G. Simon
Air Force Research Laboratory (AFRL/RXCC), 2941 Hobson Way, Wright Patterson Air Force Base, OH 45433-7750, USA

B. Antoun (ed.), *Challenges in Mechanics of Time Dependent Materials, Volume 2*,
Conference Proceedings of the Society for Experimental Mechanics Series, DOI 10.1007/978-3-319-22443-5_11

Fig. 11.1 Materials hybrid system development

yield superior mechanical and molecular connections for the entire hybrid and eliminate the need for adhesives. This paper is focused on the additive manufacturing SLM process where metal is deposited onto a carbon fiber fabric substrate with the requirement that metal must penetrate into half of the depth of one ply of carbon fabric after SLM, and the other half contains dry fibers allowing for resin infusion to build the hybrid sample later.

11.2 Experimental Methods

The Ti-6Al-4V (Ti64) powder was used for this study and was purchased from Advanced Powders and Coatings Inc. Figure 11.2 shows the particle size distribution and SEM micrograph which were provided by the manufacturer. The Ni-coated carbon fabric, satin finish, 0.5 mm thick and the un-coated carbon fabric (T300), plain weave and 0.25 mm thick, were used as the substrates in this study. The Ni-coated carbon fabric was provided by the Conductive Composites Company. The T300 fabric was provided by Fabric Development Inc.

Mount Laser Photonics Center (MLPC) in Kettering, Ohio performed the Ti64 deposit into carbon fabric using Selective Laser Melting (SLM). SLM is a powder bed additive manufacturing technique, where the sample is built layer by layer. Figure 11.3 shows the SLM schematic and close-up powder bed with fabric substrate and Ti64 powder coated fabric before sintering [2]. The experiment was done in an Argon atmosphere with an oxygen level approximate of 0.2 %. The quality of each layer build depends on proper selection of processing parameters, such as laser power, scan speed, hatch distance, and layer thickness [3–6]. In this study, hatch distance and layer thickness were maintained constant while laser power and scan speed were varied. Figure 11.4 shows the processing parameters and the volume energy density equation which was used to calculate the amount of energy transported into the material. The SLM test matrix was established by various power levels and scan speeds, and the energy density was calculated and used to discuss the quality of the deposit. A series of experiments were performed where Ti64 was deposited into the Ni-coated carbon fabric and into the un-coated carbon fabric (T300). For each deposit layer, the sample was visually observed for selections followed by examining using table top Scanning Electron Microscopy (SEM) TM3000 to determine if the Ti64/carbon depositions met the requirement. The requirement for the SLM process was that Ti64 must penetrate into half of the depth of one ply of carbon fabric, and the other half contains dry fibers allowing for resin infusion to build the hybrid sample later. One and/or two metal layers were deposited into the fabrics, and each layer required the SEM work to quantify for the requirement.

11.3 Results and Discussion

11.3.1 Results of Ti64/Nickel-Coated Carbon Fabric Deposition

Two sets of the energy density matrixes were determined: a low power with a low speed; and a high power with a high speed, as shown in Fig. 11.5, along with pictures of each sample set for a single layer deposition. Individual test sample in the test matrix had dimension of 10 mm × 10 mm and was visually examined for a decision if the sample needed further SEM examination for checking the Ti64 penetration. Recall, the requirement was that the Ti must penetrate into half of the depth of one ply of carbon fabric with the other half containing dry fibers. It was observed that when a low energy density was used (11–33 J/mm^3); the Ti64 powder did not melt. When a high energy density was used (>67 J/mm^3); the fabric substrate was destroyed (burned through). Visually, several promising depositions were observed and their energy densities were 39 and 47 J/mm^3. These samples had a shiny metal on the Ti side and a dry fiber on the fabric side. Figure 11.6 shows the SEM images of the sample deposited using an energy density of 39 J/mm^3. It was observed that the Ti64 fabric was only penetrated around 100 μm deep from the 500 μm thick fabric. Thus, increasing the energy density by higher power level is one way for further Ti64 penetration. There were three samples with energy density levels of 37, 46, and 61 J/mm^3 which were examined by SEM. The samples deposited using an energy density of 37 and 46 J/mm^3 did not show as much Ti penetration as expected

Fig. 11.2 Particle size distribution and micrograph of Ti64 powder

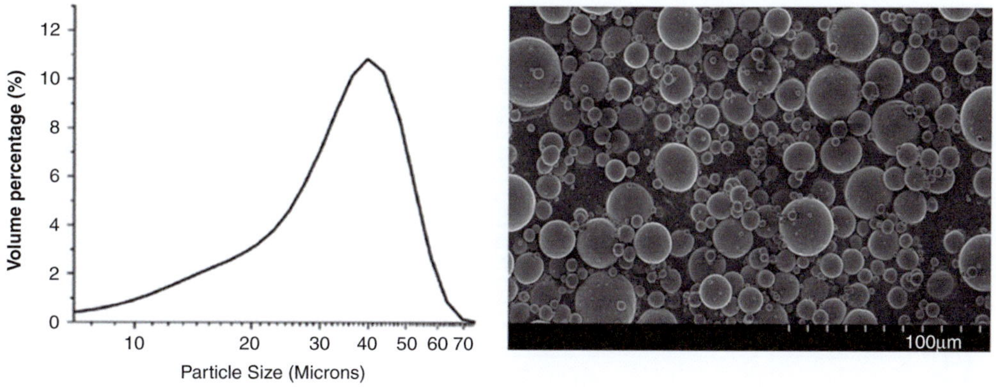

Fig. 11.3 Selective laser melting schematic and close-up view of carbon fabric and powder bed substrate

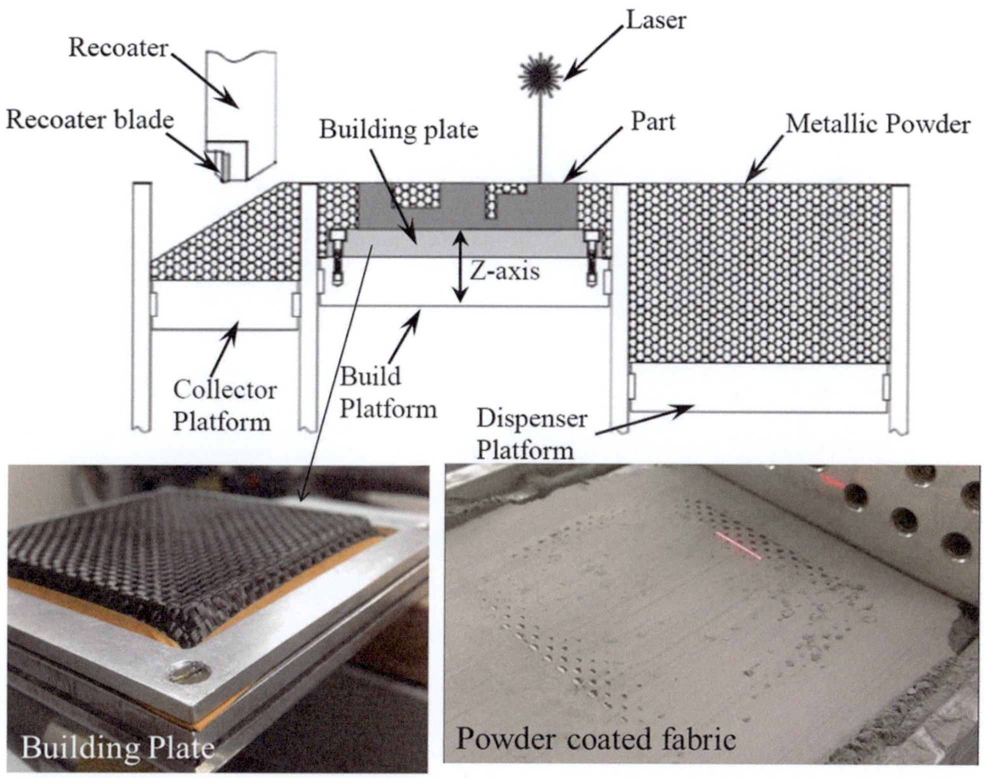

Fig. 11.4 Selective laser melting processing parameters and energy density equation

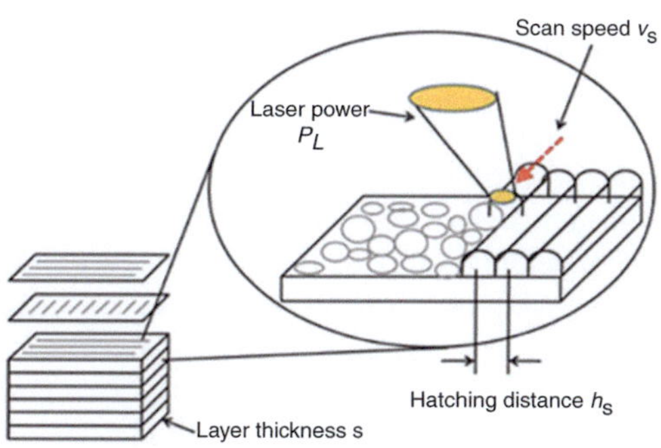

$$E_d = \frac{P_l}{v_s * h_s * s}$$

E: energy density (J/mm^3)
P_1: laser power, (W)
V_s: scan speed (mm/s)
h_s: hatching distance (mm)
s: layer thickness (mm)

Increase Scan Speed (mm/s) →
Increase Power Level (W) ↓

67	33	22	17	13	11			
100	50	33	25	20	17			
133	67	44	33	27	22			
167	83	56	42	33	28			
200	100	67	50	40	33			
233	117	78	58	47	39	29	23	19
				60	50	38	30	25
				73	61	46	37	31
				87	72	54	43	36
				100	83	63	50	42

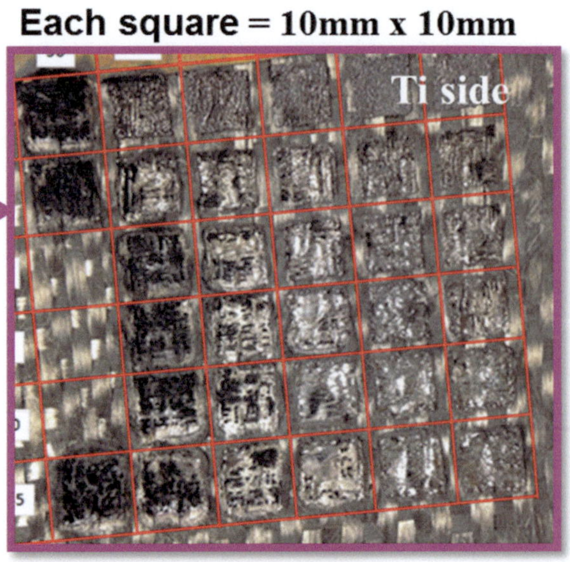

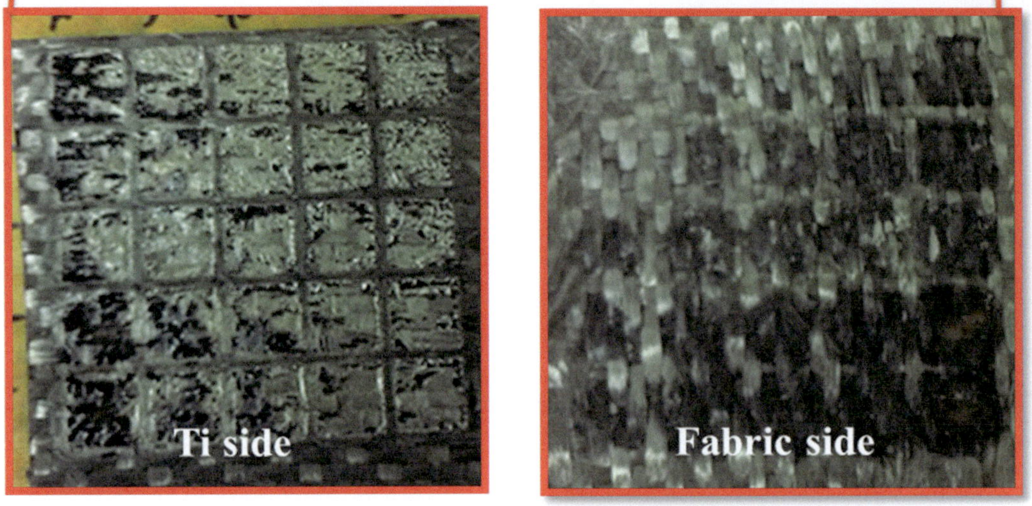

Fig. 11.5 Energy density test matrix used for Ni-coated carbon fabric substrate and images of each deposit sample

from the results above (39 J/mm³ used). The sample deposited using an energy density of 61 J/mm³ showed promise. Figure 11.7 shows the SEM images of the 61 J/mm³ sample. It was observed that in some areas, the Ti64 penetrated half of the fabric thickness and in other areas; there was little more than half of the fabric thickness saturated with Ti64 which is still within the acceptable range for the Ti64 penetration requirements.

To build a desirable metal thickness layer on top of the fabric, more than one deposit layer must be used in the deposition process. Thus, the second deposit layer processing parameters needed to be optimized. Another set of experiment was performed in which the first layer was deposited onto the fabric substrate using the energy density of 71 J/mm³, which was slightly higher than the finding value above (61 J/mm³). This was due to variations in other parameters like power, scan speed, and hatch spacing which were slightly changed for better penetration. For the second deposit layer, four different sets of parameters were used by changing the power level and scan speed while the energy density was maintained, also the hatch spacing and layer thickness were constant. Each sample was cut and examined using table top SEM. Figure 11.8 shows the images of the deposition and listed parameters used relative to the first layer. Figure 11.9 shows the SEM results of the cross-sectional areas for the sample deposited using power level (P_4) and scan speed (S_4). By decreasing both the power level and scan speed for maintaining the energy densities, the un-melted Ti64 powder was observed in the second deposit layer. The first deposited layer was similarly observed in previous work where the Ti64 penetrated half of the thicknesses onto the fabric. By increasing the power and slowing the scan speed as shown in Fig. 11.10, no Ti64 powder was observed in the second layer.

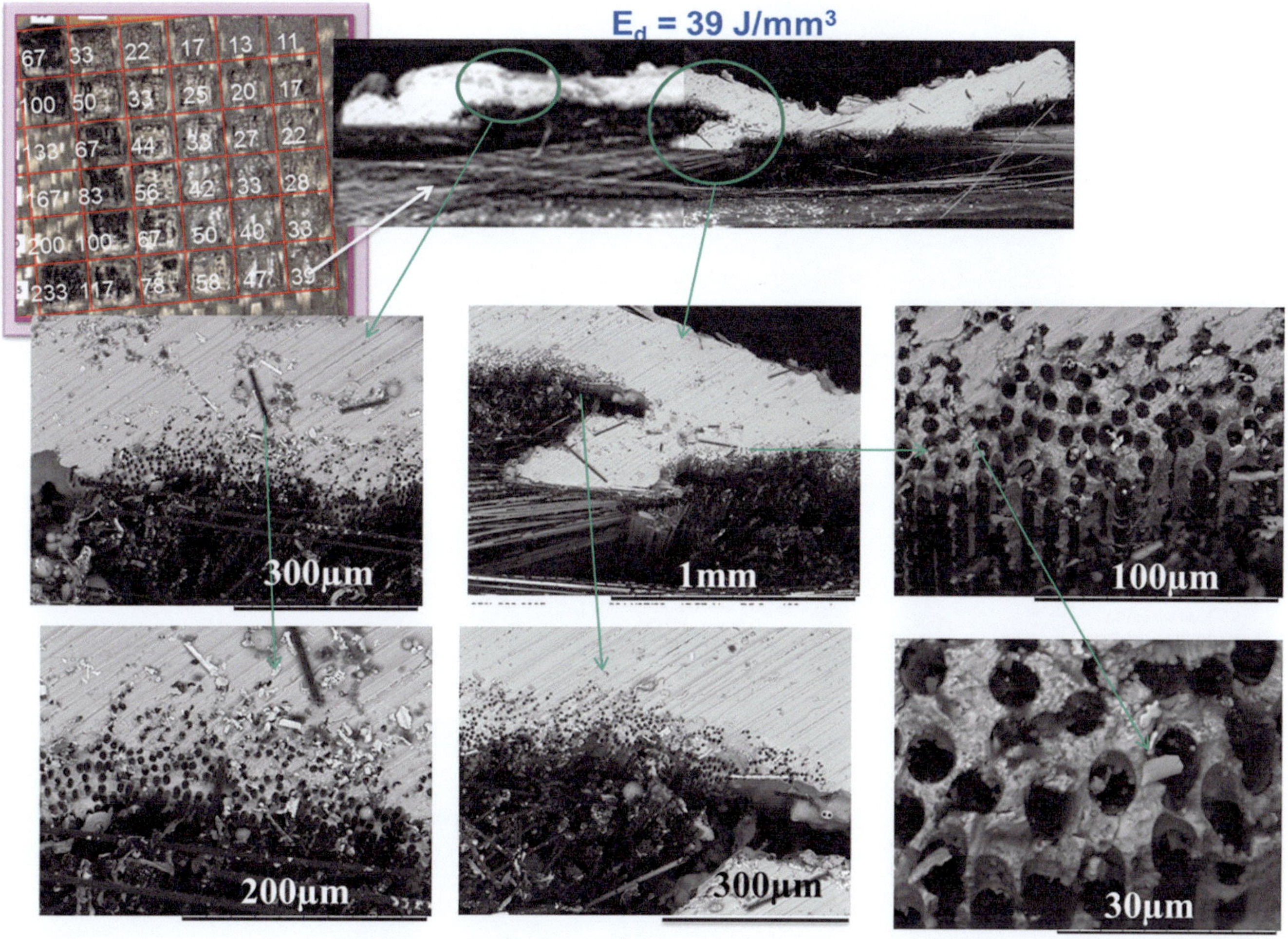

Fig. 11.6 SEM images of one Ti layer deposited into Ni-coated carbon fabric with energy density of $E_d = 39$ J/mm^3

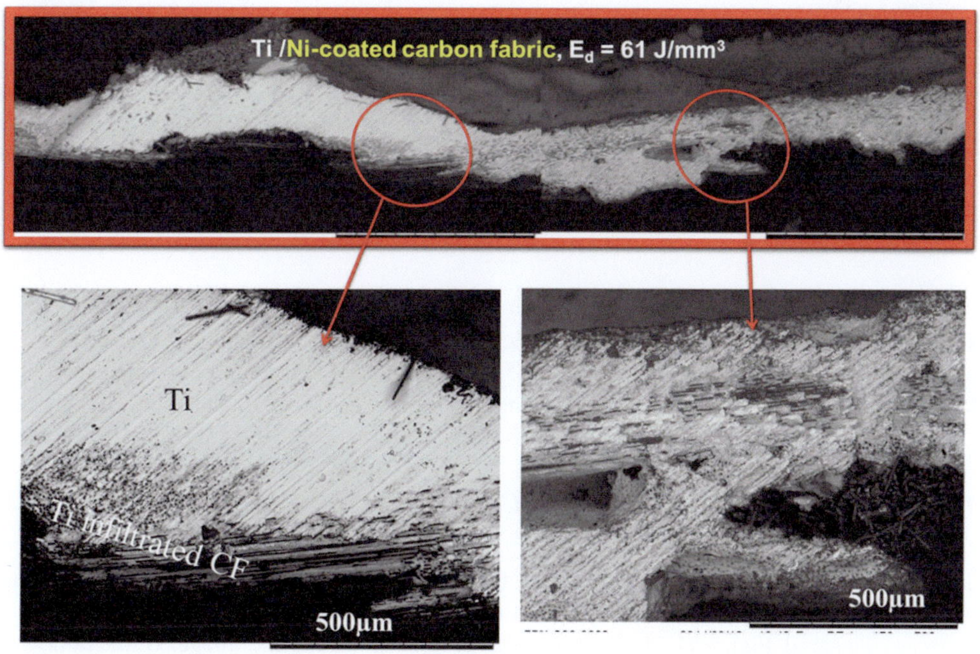

Fig. 11.7 SEM images of one Ti layer deposited into Ni-coated carbon fabric with energy density of $E_d = 61$ J/mm^3

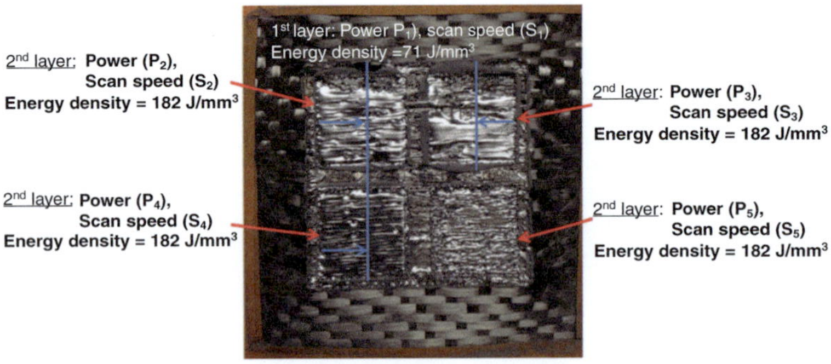

Ni–coated carbon fabric	Power (W)	Scan speed (mm/s)	Energy density (J/mm³)	SEM Obtained
1st layer	P1	S1	71	-
2nd layer-Top-Left	P2>P1	S2<S1	182	Yes
2nd layer-Top-Right	P3>P2	S1>S3>S2	182	Yes
2nd layer-Bottom-Left	P4<P1	S4<S3	182	Yes
2nd layer-Bottom-Right	P5<P4	S5<S4	182	No

Fig. 11.8 Pictures of Ti layer deposited into Ni-coated carbon fabric and their processing parameters

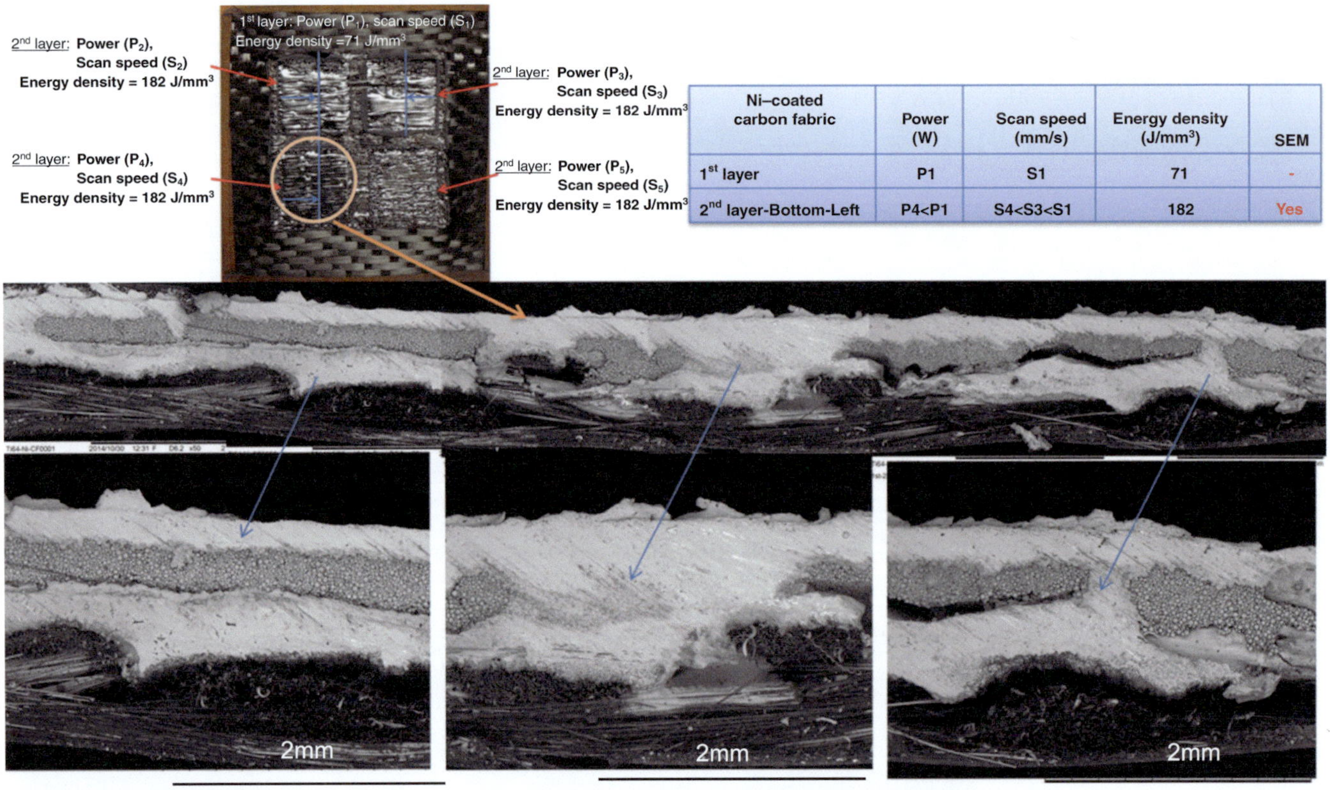

Ni–coated carbon fabric	Power (W)	Scan speed (mm/s)	Energy density (J/mm³)	SEM
1st layer	P1	S1	71	-
2nd layer-Bottom-Left	P4<P1	S4<S3<S1	182	Yes

Fig. 11.9 SEM images of two Ti layers deposited into Ni-coated carbon fabric using low power and scanning speed, results of un-melted Ti powder of second deposit layer

Ti64 penetrated into at least half of the fabric thicknesses, which met the requirement for the first layer. Thus, the processing parameters for the second deposition layer of the Ti/Ni-coated carbon fabrics were established, and the results were suggested for scale up. Scale-up for making 50 mm² samples will be performed later to process them into a structural hybrid.

11.3.2 Results of Ti64/T300 (Un-Coated) Carbon Fabric Deposition

The energy density test matrix designed for the T300 carbon fabric was similar to the energy density test matrix created for the Ni-coated carbon fabric. The deposition requirements for the T300 carbon fabric (un-coated fabric) for the first deposit layer are the same. Figure 11.11 shows the energy density values and images of the deposition. Un-like the Ni-coated carbon fabric substrate, the un-coated carbon fabric substrate behaved differently whereas the Ti64 was observed to saturate into the

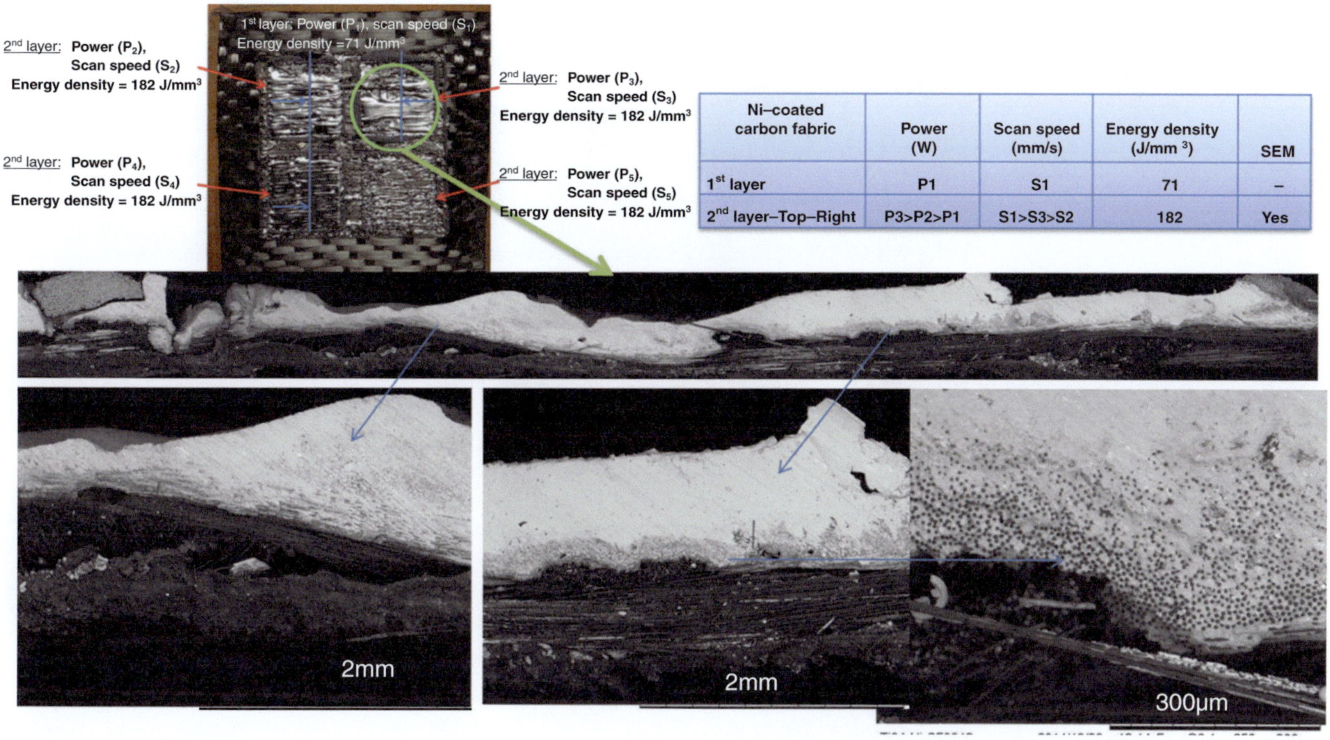

Fig. 11.10 SEM images of two Ti layers deposited into Ni-coated carbon fabric using high power and scanning speed

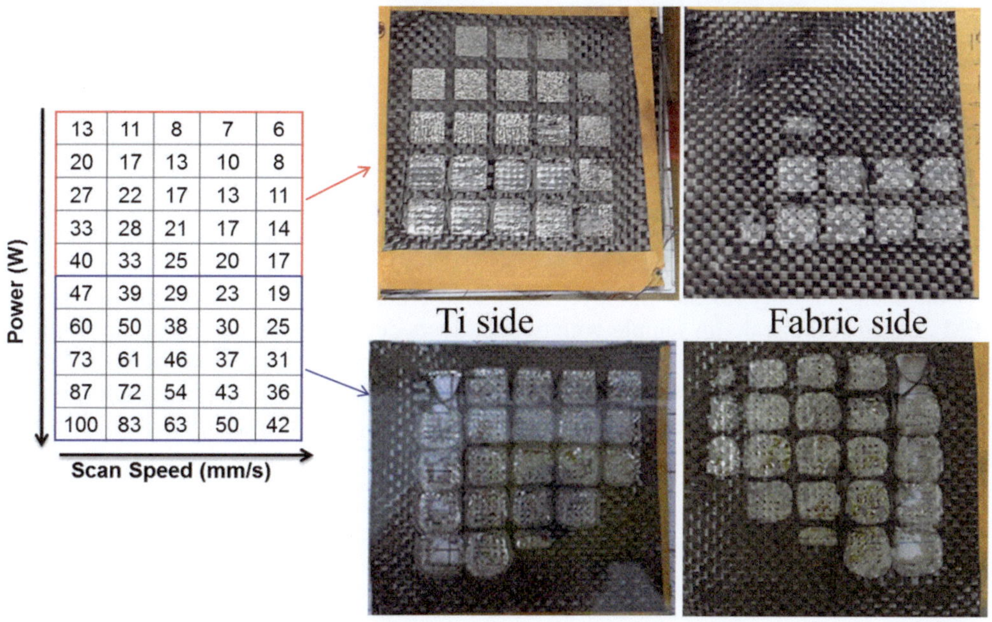

Fig. 11.11 Energy density test matrix used for T300 carbon fabric substrate and images of Ti deposited fabric (Ti side and fabric side)

T300 fabric more for the same energy density used as in the Ni-coated fabric. This was due to the uncoated fabric being thinner, and the heat absorption was different than that of the Ni-coated fabric. The samples with energy density values in the bottom half of the test matrix resulted in too much Ti saturation. Several samples in the top half of the test matrix were examined by way of cross-section analysis using SEM. Figure 11.12 shows the sample deposited using an energy density of 17 J/mm^3 as the first deposit layer. The sample showed that the Ti64 penetrated into both sides of the carbon fabric. There is a need for scale up to a large sample size utilizing the lesson learned from the Ni-coated carbon substrate above. A decision

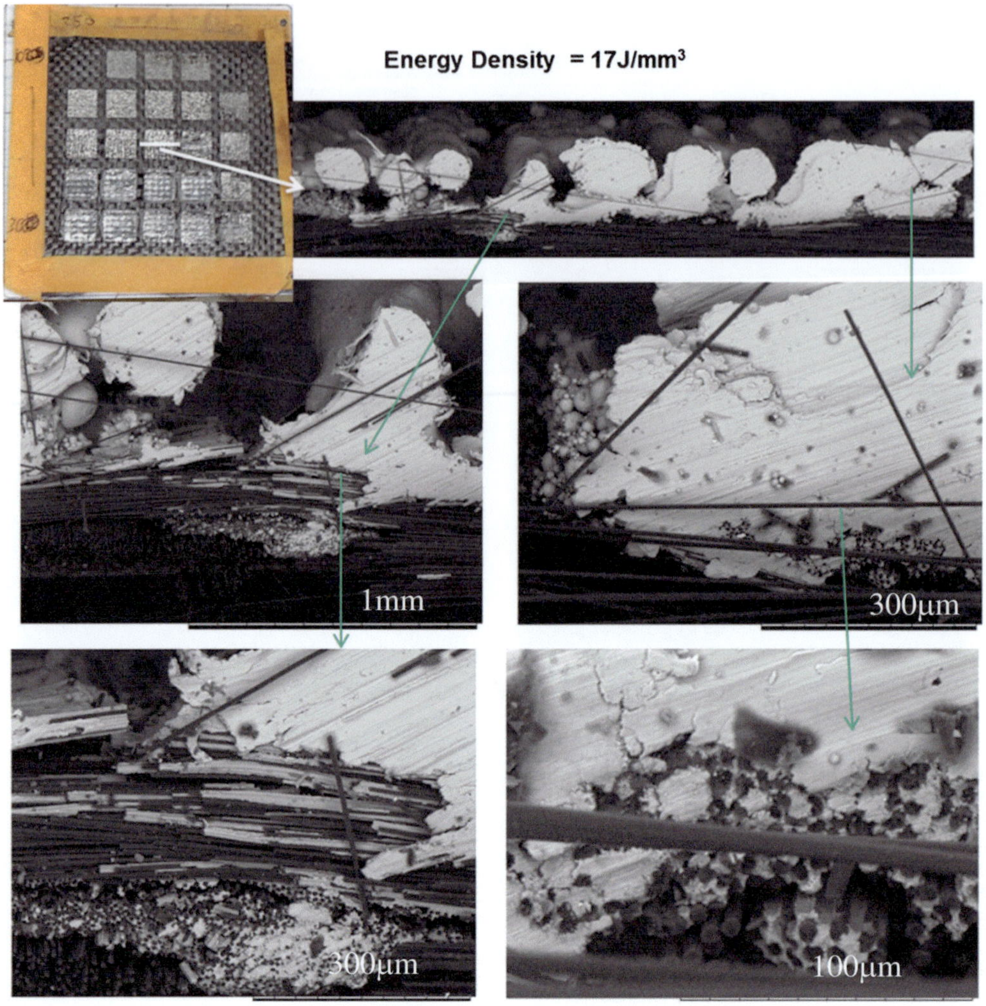

Fig. 11.12 SEM images of sample with 17 J/mm^3

was made to lower the energy density for the first layer and increase the energy density for the second layer. Figure 11.10 shows that the SEM images of the scale-up sample with two Ti layers deposited on one ply of the un-coated carbon fabric. The first layer was deposited using an energy density of 14 J/mm^3, and the second layer was deposited using an energy density of 35 J/mm^3. It was observed that the Ti64 penetrated half of the thickness (120 μm deep) into the carbon fabric in the scale up sample. We therefore believe that we have achieved a processing parameter set that could meet the requirement for the un-coated carbon fabric (T300) (Fig. 11.13).

11.4 Conclusions

A series of SLM experiments of Ti64 were performed using the Ni-coated carbon fabric and the T300 carbon fabric (un-coated) as substrates. The test samples were visually observed followed by cross-sectioning for SEM examination to determine if the depositions met the requirement. The requirement for the SLM process was that Ti64 must penetrate into half of the depth of one ply of carbon fabric, and the other half contained dry fibers allowing for resin infusion in order to build the hybrid sample later. One and/or two Ti64 layers were deposited onto the fabrics from dimension of 10 mm^2 to scale-up 50 mm^2 samples. For the Ni-coated carbon fabric, the first deposit layer met the requirement with the energy density used of 61 J/mm^3, and the second layer required a higher energy density of 181 J/mm^3 to build a desired Ti64 thickness. For the un-coated carbon fabric, both layers met the requirement in a scale-up sample where energy density of the first layer = 14 J/mm^3 and energy density of the second layer = 35 J/mm^3. The achieved processing parameters for both carbon fabric

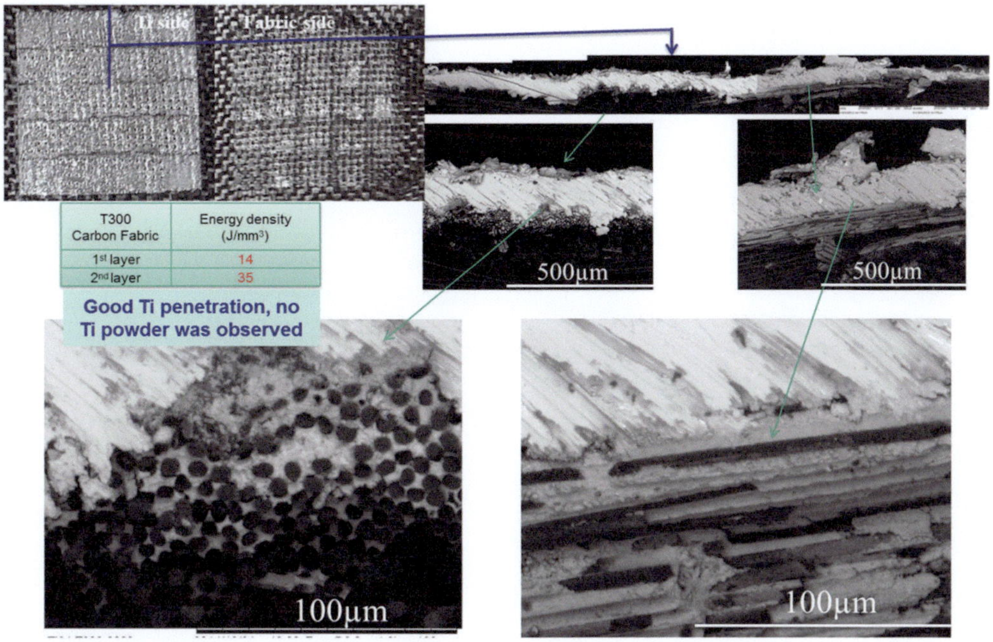

Fig. 11.13 SEM images of the scale-up sample containing two Ti layers on one ply T300 carbon fabric

substrates will be used for the scale-up to large samples for performing polymer infusion in order to create a metal/polymeric composite hybrid structure that will be used with our future work as well as with the mechanical testing of the bond strength using flatwise tension test method.

References

1. A.D. Burianeck, M.S. Spearing, Fatigue damage in Titanium-Graphite hybrid laminates. Compos. Sci. Technol. **62**(202), 607–617 (2002)
2. H. Gong, K. Rafi, T. Starr, B. Stucker, The effects of processing parameters on defect regularity in Ti-6Al-4V parts fabricated by Selective Laser Melting and Electron Beam Melting. 24th Annual International Solid Freeform Fabrication Symposium—An Additive Manufacturing Conference, Austin, TX, 2013
3. B. Vandenbroucke, J.P. Kruth, Selective Laser Melting of biocompatible metals for rapid manufacturing of medical parts. Rapid Prototyping J. (2007)
4. M. Simonelli, Y.Y. Tse, C. Tuck, Microstructure of Ti64 produced by Selective Laser Melting. J. Phys. Conf. Ser. **371**(1), 120 (2012)
5. M. Simonelli, Y.Y. Tse, C. Tuck, Further understanding of Ti-64 Selective Laser Melting using texture analysis. 23rd Annual International Solid Freeform Fabrication Symposium Austin, TX, 2012
6. A.H. Stoffregen, J. Fischer, E. Abele, Selective laser melting of porous structure. 22rd Annual International Solid Freeform Fabrication Symposium, Texas, 2011
7. K. Zeng, D. Pal, B. Stucker, A review of thermal analysis methods in laser sintering and selective laser melting. 23rd Annual International Solid Freeform Fabrication Symposium Austin, TX, 2012
8. S. Campanelli, N. Contuzzi, A. Ludovico, F. Caiazzo, F. Cardaropoli, V. Sergi, Manufacturing and characterization of Ti64 lattice components manufactured by selective laser melting. Materials **7**(6), 4803–4822 (2014)

Chapter 12
Time Dependent Response of Inconel 718

Aaron Ressa, Timothy Liutkus, Jeremy D. Seidt, and Amos Gilat

Abstract The time dependent response of precipitated hardened Inconel 718 is investigated experimentally. The experimental program is designed to determine the mechanical behavior of Inconel 718, which is extensively used in aerospace applications, at various temperatures, strain rates, and states of stress. Compression and tension tests are conducted at strain rates ranging from 0.0001 to 5000 s^{-1}. For all tests, three dimensional digital image correlation is used to measure specimen surface strains. The results show that the material is indeed sensitive to strain rate.

Keywords Inconel 718 • Rate sensitivity • Digital image correlation • Tensile Split Hopkinson Bar • Compression Split Hopkinson Bar

12.1 Introduction

Inconel is a supper nickel alloy developed for use in jet engines. It properties include high strength, good resistant to fatigue and rupture, and good mechanical properties at high temperature. These properties make it ideal for use in current jet engines used on commercial aircraft. According to FAA regulations these engines must be able to contain a turbine blade failure. To meet these requirements expensive containment tests must be conducted. To reduce the amount of physical testing, simulations are being developed with material models that rely on accurate strain rate dependence properties. High strain rate testing of precipitated hardened of Inconel 718 with a Split Hopkinson compression bar have been reported by DeMange et al. [1]. Their results show significant strain rate sensitivity that correlates well with the results presented in this paper.

12.2 Experimental Setup

Tension and compression tests are performed on Inconel 718 specimens in order to determine the effect of strain rate on the deformation and failure characteristics. These tests are conducted at strain rates of 0.0001, 0.01, 1, 500, and 2000 s^{-1}. Test specimens are fabricated from a 12.7 mm thick precipitated hardened Inconel 718 plate using wire EDM to cut the profiles. The recast layer on the surface of the specimen formed during the EDM process is ground off. The specimen gage sections are aligned in the rolling direction of the material.

The specimen gage section measures 0.762 mm thick, 2.032 mm wide, and 5.08 mm length. The compression tests are performed on a cylinder specimen 3.81 mm diameter and 3.81 mm long. The specimen ends are lubricated and compressed between two rigid tungsten carbide tips. Drawings for these specimens are shown in Fig. 12.1.

The 0.0001, 0.01, and 1 s^{-1} strain rate tests are conducted on a servo-hydraulic load frame. For the tension tests, flat serrated wedge grips grip the specimen. A digital image correlation (DIC) system is used to make full field strain measurements of both the tension and compression specimens. The specimens are prepared with a black and white speckle pattern consisting of a thin white base coat and black contrast pattern using spray paint. Two Point Grey Research GLZ-41C6M-C cameras with a 2048 × 2048 pixel resolution are focused on the specimen gage section with a 15° angle between them to form a stereographic image using the speckle pattern. Images are taken during the test and processed with Correlated Solutions VIC-3D 7 software. The Henky strain is then calculated using the surface displacements. A virtual extensometer, as shown if Fig. 12.2, is used to measure the relative displacement of the entire gage section.

A. Ressa (✉) • T. Liutkus • J.D. Seidt • A. Gilat
Department of Mechanical Engineering, The Ohio State University, 201 W. 19th Ave., Columbus, OH 43210, USA
e-mail: ressa.8@osu.edu

B. Antoun (ed.), *Challenges in Mechanics of Time Dependent Materials, Volume 2*,
Conference Proceedings of the Society for Experimental Mechanics Series, DOI 10.1007/978-3-319-22443-5_12

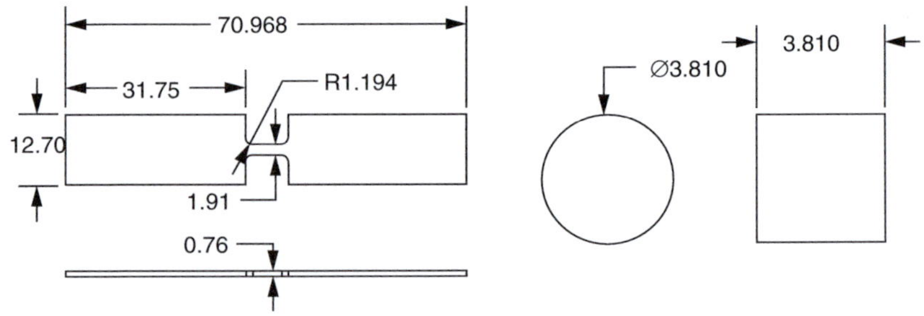

Fig. 12.1 (**a**) Tension Specimen (**b**) Compression Specimen

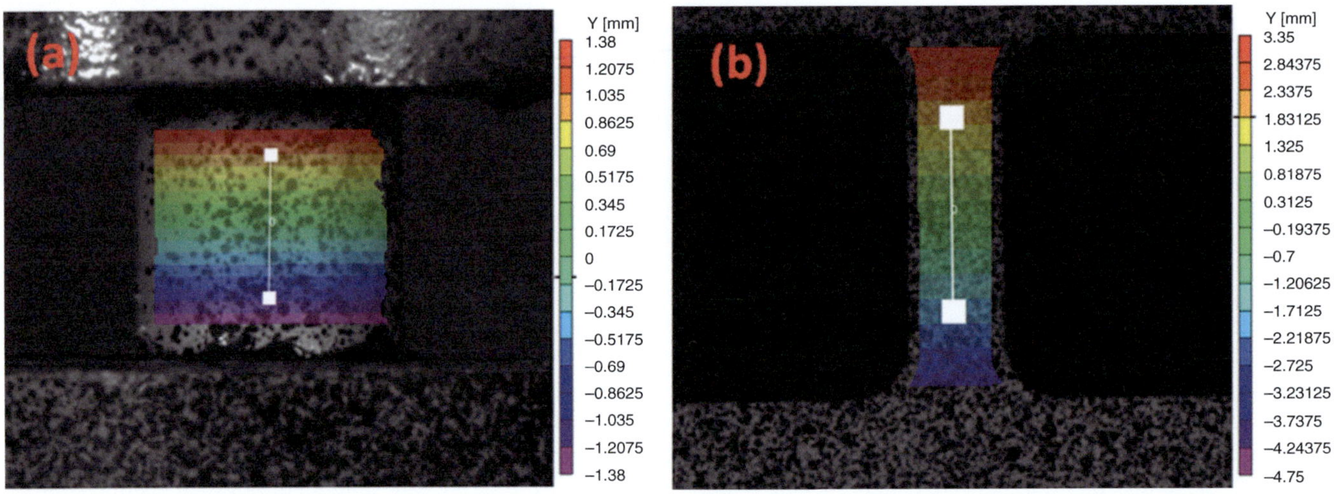

Fig. 12.2 (**a**) Compression DIC results with 2 mm extensometer. (**b**) Tension DIC results with 4 mm extensometer

Tension tests at strain rates of 500 and 2200 s^{-1} are performed on a 7075 aluminum, 12.7 mm diameter tension Split Hopkins bar consisting of a 3.68 m long incident bar and a 1.8 m long transmitter bar. This test setup provides a loading wave of 500 μs. Analysis of the waves measured in the incident and transmitter bars is used to determine the force, strain rate, and strain in the specimen. The specimen is attached to each the bars with an adapter bonding the specimen flange to the end of the incident and transmitter bar. A pair of Photron SAS.1 high speed cameras is used to make full field 3D surface displacement measurements. The frame rate of the cameras determined depending on the speed of the test, ranging from 100,000 to 250,000 frames per second.

Compression tests at strain rates of 1000, 2100, and 5200 s^{-1} are conducted on a 12.7 mm diameter, Ti-6AL-4V compression split Hopkins bar (SHB). The compression SHB setup, shown in Fig. 12.3, consists of a striker bar that contacts a incident bar, generating a 250 μs length loading wave. The test conducted at strain rate of 5200 s^{-1} uses a compression specimen measuring 3.81 mm in diameter and 2.032 mm in height. The ends of the specimen are lubricated and placed between the incident and transmitter bars. The specimen contacts a threaded tungsten carbide insert on the tip of each bar. 3D DIC measurements are made with the same system used in the tension test.

12.3 Results

Figure 12.4 shows strains measured in tensile SHB tests. The figure show strains determined from the wave, the strain measured directly on the specimen using a 4 mm virtual extensometer, and the strain measured with DIC at the failure point. The three strain measurements agree until the necking point. The local strain at the failure point increases as the specimen necks. This strain increase in not seen in the strain determined from the wave or from the 4 mm virtual DIC extensometer. This result can better help define plastic behavior and fracture models for use in finite element analysis.

Stress strain curves for all tension tests are shown in Fig. 12.5. The results show that the yield point is dependent on strain rate. The hardening curve is not affected by the strain rate.

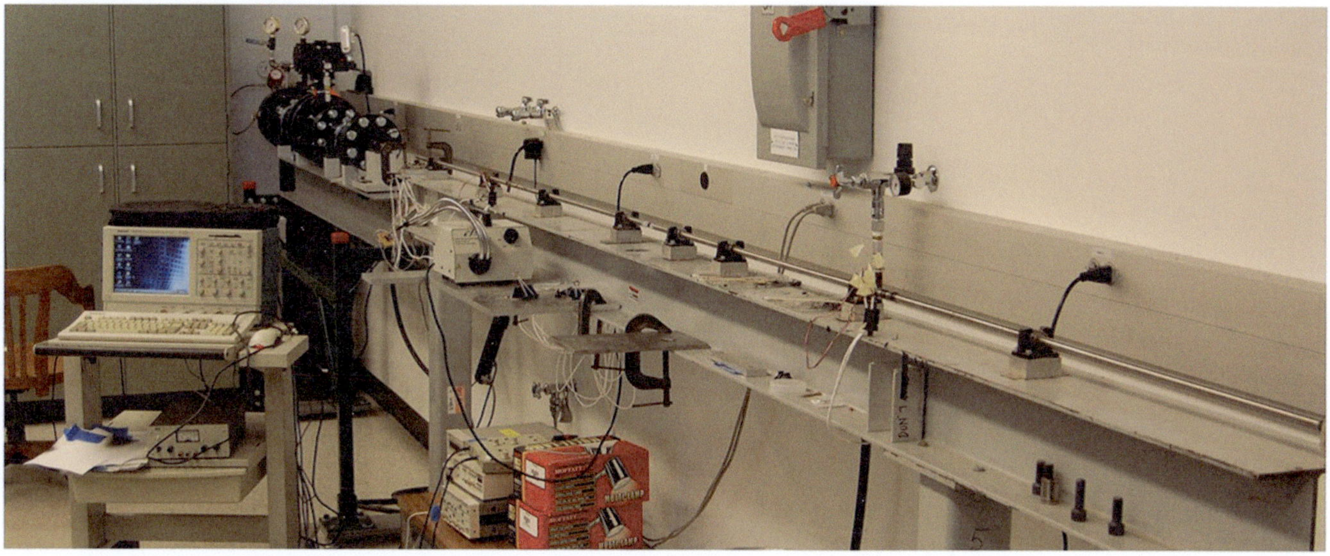

Fig. 12.3 Compression Slit-Hopkinson Bar test apparatus with wave recorder

Fig. 12.4 Strain measurements versus time. The wave data is plotted with strain of the virtual strain gage and that of the failure point

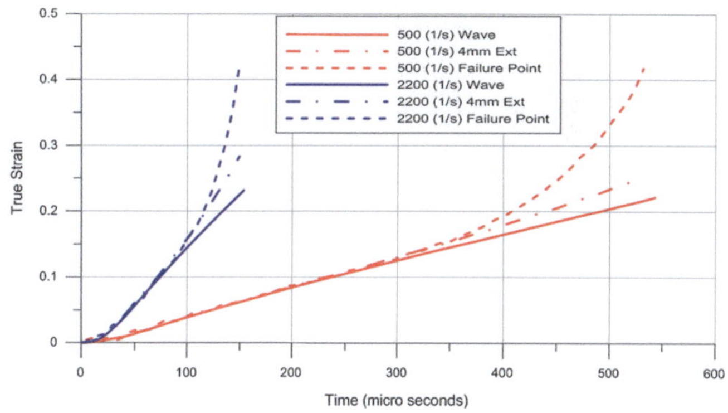

Fig. 12.5 True strain vs. true stress at the various strain rates tested in tension

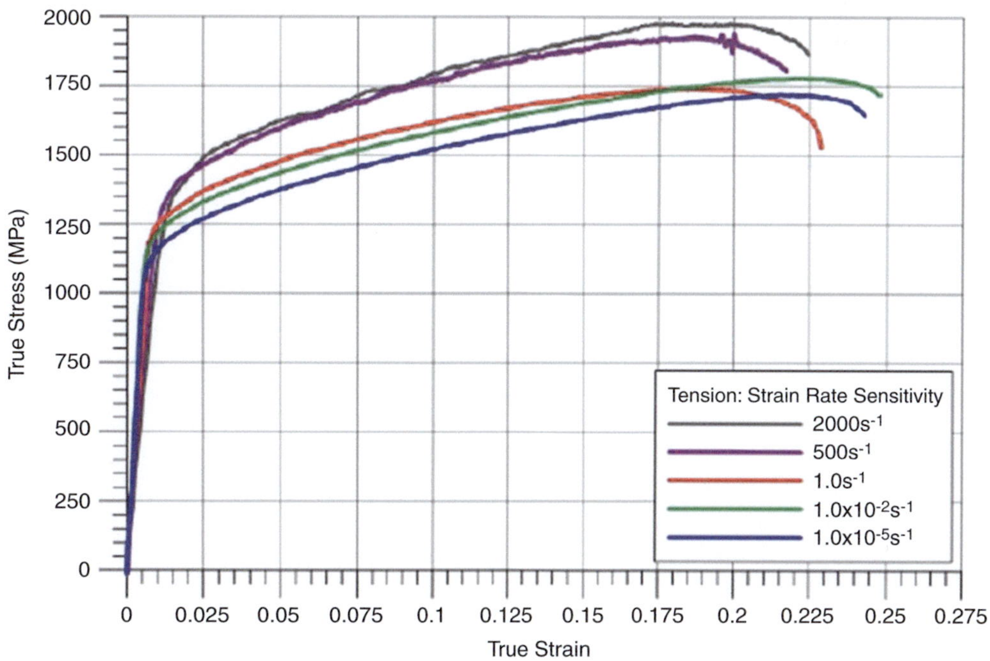

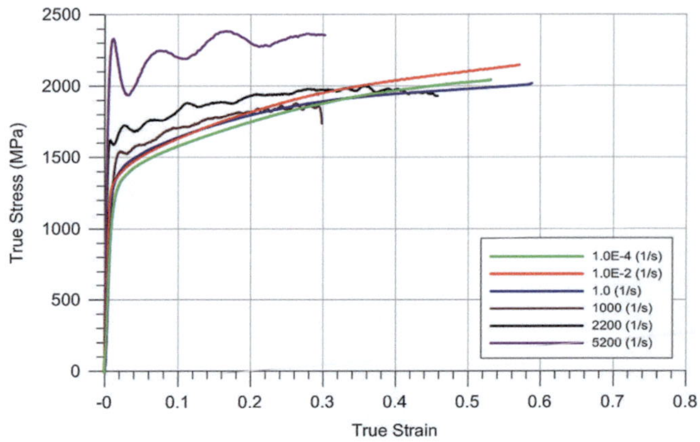

Fig. 12.6 True stress vs. true strain at various strain rates in compression

The compression test results, plotted in Fig. 12.6, also show an increase in stress levels as the strain rate increases. The stress levels measured in this paper are similar to the results found by DeMange [1]. The results in this paper show a decrease in the hardening as the strain rate increase, a trend not seen in DeMange [1] results.

12.4 Summary and Conclusion

Detailed results from the testing of precipitated hardened Inconel 718 at a variety of strain rates ranging from 0.0001 to 2200 s^{-1} in tension, and 5200 s^{-1} in compression. The results show the Inconel 718 is rate sensitive in both tension and compression. The use of digital image correlation produces more accurate results and should be incorporated in testing procedures used to define material models in order to provide more accurate hardening and fracture data.

Acknowledgement The research was supported by the USA Federal Aviation Administration, Grant 11-G-003.

Reference

1. J.J. Demange, V. Prakash, J.M. Pereira, Effects of material microstructure on blunt projectile penetration of a nickel-based super alloy. Int. J. Impact Eng. **36**(8), 1027–1043 (2009)

Correction to: Effects of Net and Solid Skins on Self-Supporting Lattice Structures

A.O. Aremu, Ian Maskery, C.J. Tuck, I.A. Ashcroft, R.D. Wildman, and R.J.M. Hague

Correction to:
Chapter 10 in: B. Antoun (ed.), *Challenges in Mechanics of Time Dependent Materials, Volume 2*, Conference Proceedings of the Society for Experimental Mechanics Series, https://doi.org/10.1007/978-3-319-22443-5_10

This book was inadvertently published with incorrect initial for one of the authors of Chapter 10, which has now been updated.

The author, Dr. Ian Maskery, does not have a middle initial as found in the book chapter. The chapter contact author and volume editor added the middle initial "A," but it is not part of his name and impacts his ORCHID details.

The updated version of this chapter can be found at
https://doi.org/10.1007/978-3-319-22443-5_10

B. Antoun (ed.), *Challenges in Mechanics of Time Dependent Materials, Volume 2*,
Conference Proceedings of the Society for Experimental Mechanics Series, https://doi.org/10.1007/978-3-319-22443-5_13

MIX
Papier aus verantwortungsvollen Quellen
Paper from responsible sources
FSC® C105338

If you have any concerns about our products,
you can contact us on
ProductSafety@springernature.com

In case Publisher is established outside the EU,
the EU authorized representative is:
Springer Nature Customer Service Center GmbH
Europaplatz 3, 69115 Heidelberg, Germany

Printed by Libri Plureos GmbH
in Hamburg, Germany